爱屋及邬
Love Home　Love Hudec

纪念邬达克绘画雕塑邀请展
Painting and Sculpture Invitational Exhibition in Memory of L. E. Hudec

李向阳 主编
Editor: Li Xiangyang

上海远东出版社

爱屋·及邬
—— 纪念邬达克绘画雕塑邀请展

编委会顾问	胡劲军
艺术顾问	毛时安
编委会主任	褚晓波

编　委　会（按姓氏笔画为序）
毛时安　史晓宛　朱　光　刘素华
李孔三　李向阳　陈　梁　陈振民
胡劲军　侯斌超　贺寿昌　徐海清
高文虹　章和轼　褚晓波　葛乾巽
舒晟岚　谭玉峰

主　编	李向阳
责　编	贺　寅
翻　译	李应登
	王昌玲
摄　影	林劲松
装　帧	曹景宇

Love Home, Love Hudec
—— Painting and Sculpture Invitational Exhibition in Memory of L. E. Hudec

Editorial Advisor：Hu Jinjun
Artistic Advisor：Mao Shi'an
Editorial Director：Chu Xiaobo

Editorial Board(in alphabetic order)

Chen Liang	Chen Zhenmin	Chu Xiaobo	Gao Wenhong
Ge Qianxun	He Shouchan	Hou Binchao	Hu Jinjun
Li Kongsan	Li Xiangyang	Liu Suhua	Mao Shi'an
Shi Xiaowan	Shu Shenglan	Tan Yufeng	Xu Haiqing
Zhang Heshi	Zhu Guang		

Editor in Chief：　Li Xiangyang
Editor：　　　　　He Yin
Translators：　　Li Yingdeng　Wang Changling
Photographer：　Lin Jinsong
Cover Designer：　Cao Jingyu

"爱屋"是本能、是欲望　　"及乌"是教养、是情怀

"To love one's own home" is just one man's nature and instinct while, "To love its environs" is nurture and intelligent.

序 一

建筑大师贝聿铭曾经说过，建筑是一种社会艺术形式。对于上海这座历史文化名城而言，建筑不仅承载着城市的历史记忆，更是城市文脉延续的重要载体。在市委市府的重视之下，上海文化艺术产业迅猛发展，越来越多的跨界项目在城市的各个角落应运而生，以海派建筑文化保护和传承为使命的"爱屋·及邬"纪念邬达克绘画雕塑邀请展，在12月上海国家历史文化名城30周年宣传之际推出，具有非常的意义。

展览汇聚沪上国画、油画、水彩、雕塑多种艺术形式的艺术家，他们用不同的艺术手段和视角去捕捉邬达克建筑与上海这座城市的关系。参展的艺术家们都是近些年都市与建筑题材创作成果斐然的艺术家，他们深入上海的历史风貌街区，用不同的笔触表达着他们对城市记忆、城市发展的思考。如果说邬达克的建筑遗产是二十世纪上半叶，上海这块海纳百川土地上中西方文化交融的硕果，那么"爱屋·及邬"受邀参展艺术家的创作，则是一次跨世纪的建筑艺术与绘画、雕塑艺术的融合，更是当代艺术家们对"创新"、"融合"的现代城市精神的一次诠释。

我们欣喜地看到，组织这样活动的主体，不仅是政府机构，还有像邬达克文化发展中心这样的社会组织。近几年来，他们从修缮邬达克旧居做起，创建邬达克纪念馆、举办各类主题文化活动，成为城市文化传承和传播的使者。在国家鼓励文化、文物单位与社会力量深度合作的大背景下，他们用"爱屋及乌"般的热情与努力，点燃上海民间力量参与历史文化遗产保护与传承的薪火，感染了越来越多的人们走出家门，拿起相机、执起画笔，探寻属于他们自己心中的"邬达克"，表达着他们对上海这座不断前行的现代化城市历史与未来的关注。

胡劲军
中共上海市委宣传部副部长
2016年9月

PREFACE ONE

I. M. Pei, the renowned architect once remarked, "Architecture is a form of social art." For Shanghai, a city well-known for its long history and kaleidoscope culture, architecture is not only the carrier of historic memories of Shanghai, but also the important medium of cultural continuity as well. Shanghai Municipal Government and Municipal Party Committees contribute a lot to the promotion of the robust development of cultural and art sectors. As a result, more and more cross-boundary projects come out in each corner of Shanghai. And the very significant one is "Love Home, Love Hudec" –A Painting and Sculpture Invitational Exhibition in Memory of L. E. Hudec to be held to publicize the 30th Anniversary of Shanghai National Historic and Cultural City in December 2016, taking the mission to protect and preserve Shanghai architectural culture for the future generations.

The Exhibition attracts artists in various sectors—traditional Chinese painting, oil painting, watercolors and sculptures. They've captured the interrelationship between Hudec's architecture and Shanghai in diversified artistic ways and from distinct perspectives. All the invited artists have masterpieces on metropolis and architecture. They have close observation of historic blocks and express their thoughts on the memory and the development of Shanghai in various styles. If we take Hudec's architectural heritage as a brilliant success of the integration of western culture with Chinese culture that occurred in the broad bosom of Shanghai in the first half of the 20th century, then the works of the invited artists in this exhibition shall be taken as a trans-century convergence of Hudec's architecture with painting and sculpture, or rather an embodiment of modern city spirit — "Innovation" and "Convergence".

We are delighted that apart from governmental efforts, some non-government organizations like Hudec Culture Development Center also begin to host such events. In recent years, they become an emissary to transmit and propagate the urban culture starting from renovating Hudec's Residence, establishing Hudec Memorial Hall to hosting various cultural events. The State Government of China proactively advocates the protection of cultural heritages, and the cooperation between Administration for Protection of Cultural Relics and non-government organizations goes further. Under this circumstance, "Love Hudec, Love Home" is held with great passion and exertion to ignite the civilian power to protect and transmit historic and cultural heritage, inspiring more and more people to go out with cameras and painting brushes to search Hudec in their heart and express their concern on the past and the future of Shanghai, an ever-renewing modern metropolis in China.

By Hu Jingjun
Vice Director of Publicity Department of
CCP Shanghai Committee
September, 2016

序 二

上海是一座有着悠久历史的现代化国际大都市，拥有丰富的文化遗产资源。1986年12月8日，上海被国务院公布为国家历史文化名城。作为中国近代百年史中具有特殊地位的城市，近代上海不仅是中国的商业中心、贸易中心和金融中心，也是全国建筑活动最为频繁、成就最为突出的城市。上海自1843年开埠后，各国将不同建筑风格的教堂、银行、办公楼、旅馆、影剧院、商店、医院、花园住宅、公寓等西方建筑形式引入上海并大批建造，数量之巨、质量之精是举世公认的。上海的近代建筑是世界建筑史和中国建筑史的重要组成部分，更是上海弥足珍贵的历史文化遗产，这也构成了上海这座城市别具一格的品质和特性。

在上海众多的优秀近代建筑中，仍然较完整地保存了许多邬达克建筑。今天，这些建筑已成为上海城市风格的重要组成部分。邬达克在上海创造的建筑传奇，既是东、西方文化在上海融合萃取发展的见证，也是上海"海纳百川、追求卓越、开明睿智、大气谦和"城市精神的体现，更是上海人共同的城市记忆与乡愁。

时值上海被国务院公布为国家历史文化名城30周年之际，第二届邬达克建筑遗产文化月如期举行。此次由邬达克文化发展中心组织发起的展览主题是"爱屋·及邬"，一个由"爱"出发纪念邬达克的绘画与雕塑艺术展。策展及参展艺术家都来自上海，他们对生活的这座城市历史与发展有着长期的关注与创作实践，参展的作品更是表达了艺术家们心中的"乡愁"。可以说，"爱屋·及邬"是上海艺术家、学者用色彩、笔触、造型、文字对上海这座高速发展的国际大都市的一次深刻历史回望与未来思考，也是城市建筑遗产与当代艺术家跨界融合的一次有益尝试。

褚晓波
上海市文化广播影视管理局、上海市文物局副局长
2016年10月27日

PREFACE TWO

As a modern international metropolis with a long history, Shanghai is abundant in cultural heritage. On Dec. 8th, 1986, Shanghai was granted as "State-listed Historical and Cultural City" by the State Council. Shanghai, a city with special status in the century-old modern history of China, is not only the national center of commerce, trade and finance, but also the destination of architecture that witnessed the most construction projects and the greatest architectural achievements throughout China. Since Shanghai was opened as an international port in 1843, many forms of architecture in various styles from various countries have been mass imported to China, such as churches, banks, office buildings, hotels, cinemas, theatres, department stores, hospitals, villas and apartments, all of insurmountable quality. The modern architecture of Shanghai belongs to the history of Chinese architecture and that of world architecture as well, making the most precious historic and cultural heritage for Shanghai. It is the modern architecture that gives a unique charm and personality to Shanghai.

Among myriads of excellent modern buildings in Shanghai, almost all those designed by Hudec are well preserved. Today, these buildings have become an essential part of Shanghai landscape. The legend of architecture that Hudec created here represents the convergence of the Oriental and Western cultures in Shanghai, embodies Shanghai's characteristics —"Broad and Open, Wise and Smart, Grand and Humble", and carries the memory and the nostalgia of all Shanghainese.

December 8th, 2016 marks the 30th anniversary of Shanghai being classified as "State-listed Historical and Cultural City" and the 2nd Hudec Architectural Heritage Cultural Month will be held as scheduled. The theme for this exhibition that Hudec Cultural Development Center initiated is "Love Home, Love Hudec". It is a painting and sculpture invitational exhibition in memory of Hudec that is totally out of LOVE and for LOVE. The curator and the invited artists are all from Shanghai. They are well-known in art circles for their long time concern about and creation on the history and development of the city where they live. The works on exhibition express the "nostalgia" as these artists have felt. To some extent, "Love Home, Love Hudec" demonstrates a recall of the past and an outlook of the future expressed via colors, lines, shapes and letters for Shanghai, an international metropolis in rapid development. Besides, "Love Home, Love Hudec" can also be counted as a tentative exploration of integrating urban architectural heritage with modern art.

By Chu Xiaobo
Shanghai Municipal Administration of Culture, Radio, Film and Television
Vice Director of Shanghai Administration for Protection of Cultural Relics
On Oct. 27th, 2016

序 三

鸟儿停在这里

蓝天下,总可以看见有几只小鸟,它们展翅欲飞,却永远停在了邬达克旧居高高耸立的红色尖顶上。倦鸟思归,家、屋、巢,是肉体永远的归宿,也是精神永久的所爱。爱,可以蔓延,可以想象,于是,中国人就有了"爱屋及乌"的独特诗意和浪漫。再延伸,就有了当下艺术家们"爱屋·及邬"——一个那么富于汉字美丽想象的艺术创意。邬达克,一个把自己所有才华都留在了异邦他乡,一个书写过上海城市建筑美好时光的欧洲建筑家。在这座城市生活的每一个人都会在自己人生旅途中不期而遇地邂逅打着邬达克印记的大大小小的"屋子"。

艺术家们的目光像小鸟,在我们的城市上空飞翔盘旋,然后,落在了邬达克的"屋子"上了。然后,他们把色彩尽情倾泻在画布和宣纸上。再然后,他们把自己深深的爱和敬意献给了这位消失在历史烟尘里的建筑艺术家。爱屋及邬,一切都来得那么的自然,那么的水到渠成。

时间永远在流动,但它可以凝固。凝固在建筑里,凝固在艺术里。而胶合这种凝固的材料就是"爱"。我深信,每一个参观者都会沉浸在邬达克和艺术家们"爱"的暖意中,为之深深感动。

"月明星稀,乌鹊南飞。绕树三匝,何枝可依?"爱,就是精神的屋子。我们是飞翔在星空下的小鸟,就,停在这里。

毛时安
中国文艺评论家协会副主席
上海市政府参事室参事
上海美术家协会理论委员会主任
2016 年 10 月 10 日

PREFACE THREE

Birds Linger Right Here

Under the clear blue sky, several birds are always visible, stretching their wings, ready to fly, but lingering forever there at the red ridge high above Hudec's Residence. When birds are tired, they will miss their nests. Our family, home or nest is where we dwell physically, and is also the place our heart is always longing for. Love, can extend, and can associate. So, we have a Chinese idiom - 爱屋及乌 , literally meaning "Love home and love birds perching on its roof" in English, which is uniquely poetic and romantic. If love extends further, then there comes another phrase- 爱屋及邬 , meaning "Love Home, Love Hudec" created by contemporary artists in imitation of the Chinese idiom " 爱屋及乌 ". What an artistic creation! How expressive those Chinese characters are! Hudec, a European architect, devoted all his talents to beautifying Shanghai in a foreign land far away from his homeland, and wrote down the most wonderful moments of urban architecture in Shanghai. Every Shanghai dweller in his journey through life will encounter a "home" designed by Hudec, be it short or tall, small or big.

Artists' searching eyes are like birds that circle over our city, and then alight upon Hudec's "home". Then, they pour all the colors and emotions onto the canvas and the Xuan paper. And then, they show their deep love and respect for Hudec, a late architect and artist who will be recorded by history. "Love Home, Love Hudec" — it all just happens naturally.

Time flows on forever, but it can be bound in architecture, or in art. And the binding material is "love". I am in full conviction that each visitor will be immersed in the warmth of "love" from Hudec and the artists, and deeply touched as well.

Stars are scarce and the moon is bright.
A flock heading south, tired of flight,
Circle around a tree, pondering,
Which high branch to perch tonight?
Love is the home where our spirit dwells. We are the birds flying in the starry sky before we linger right here.

By Mao Shi'an
Vice-chairman of China Literary and Art Critics Association
Counselor of Shanghai Municipal Government
Director of Theory Committee, Shanghai Artists Association
On Oct. 10th, 2016

目录 CATALOGUE

序 一 ／胡劲军 Preface One ／ Hu Jingjun	6
序 二 ／褚晓波 Preface Two ／ Chu Xiaobo	8
序 三 ／毛时安 Preface Three ／ Mao Shi'an	10
《爱屋·及邬》主题阐述 ／李向阳 The Curator's Remarks ／ Li Xiangyang	17
结缘上海 追求摩登 ／徐海清 Pursuit of Modernity in Shanghai ／ Xu Haiqing	22
杜海军 Du Haijun	42
李乾煜 Li Qianyu	54
有故人的城池才是故乡 ／朱光 Hometown is the place that housed our family and old friends ／ Zhu Guang	66
洪　健 Hong Jian	72
毛冬华 Mao Donghua	84
应海海 Ying Haihai	98
为一座城市造房子的人 ／毛时安 The man who built homes for a city ／ Mao Shi'an	112
张安朴 Zhang Anpu	122
陈　键 Chen Jian	134
贺寿昌 He Shouchang	146
后记 ／刘素华 Afterword ／ Liu Suhua	159

《爱屋·及邬》主题阐述
THE CURATOR'S REMARKS

《爱屋·及邬》主题阐述

李向阳

番禺路129号深处,有一幢英国都铎风格的建筑,立面凹凸有致,内饰完美精良。尽管花园不复存在,天际线也被四周林立的高楼大厦所淹没,但西面山墙上的鸽舍依然醒目。几只被悉心安置在栖息台上的鸽子,虽不会飞,却真假难辨,与屋顶的飞鸟雕塑前后顾盼,动静相宜,一派爱屋及乌的温暖景象。这是修葺一新的传奇人物邬达克当年为自己营造的家。

爱屋之心,人皆有之,鸥鸟都有自己的巢。与其他文明圈里的人相比,我们的同胞尽管不太喜欢自己动手修瓦补墙剪枝割草,但在表达这份爱意时却显得更幽默、更透彻。夫妻可以离婚,兄弟可以翻脸,即便搭上儿孙福利仕途前程,也义无反顾、在所不辞。然及乌者,好像凤毛麟角。强拆时有发生,渣土还在飞扬,确保自家屋中的文明秩序,出门遵守好"七不"规范,就是对当下社会的积极贡献。所以,从某种意义上说,"爱屋及乌"是一种有待普及的信仰,"爱屋"是本能,是欲望,而"及乌"是教养,是情怀。

为我们点燃这种爱、物化这种爱、并赋予其生命和美学的,当然是建筑师。没有什么东西可以象建筑那样,呵护着我们的生老病死,包裹起我们的爱恨情仇。不同的建筑师为我们设计了不同的生活,不同的生活形成了我们各自的家园。毫无疑问,邬达克是一位杰出的建筑师。他的成功,不仅仅表现为为上海留下了60多幢包括酒店、住宅、教堂、医院、影院在内的优秀历史建筑,使上海之所以成为了上海;也不仅仅体现在这些建筑的风格迥异、筑造技术的驾轻就熟,创造了一个人与一座城市建筑史的神话;而是隐含其中的一种对业主、对政策、对环境、对潮流的洞悉和把握,一种对不同文化的理解和贯通,一种遵循天意、顺其自然、知命而行的豁达与从容。今天,我们怀念邬达克,是因为我们向往昨日的气质与格调;我们追寻邬达克,是因为我们忘不掉在这些熟悉的屋宇中曾经发生过的刻骨铭心、感同身受的故事和回忆。

艺术家是另一种建筑师,为我们构筑着精神家园。他们和建筑师的视角或许不同,但同样关注着我们的生活状态,启发我们不断地对人与世界的终极关系发问。现代文明将整个世界格式化之后,我们原有的生活方式也被打乱,千城一面、日益趋同。那些从一样的混凝土森林中走出来的人,何以面对家园,何以坚守独立的精神和品格,貌似少了参照和空间。许多年来,在迅猛的都市化进程中,我们的肉身走得太快,灵魂却落在了后面,光鲜而空虚着、鸡血并孤独着。一如建筑师为我们遮风挡雨那样,艺术家以其敏锐而炽热的笔触,剥开我们的躯壳,直抵我们的内心,帮我们安顿走失的灵魂。他们画建筑,不一定关注建筑美学的讨论,画街景,也不是对城市肌理的研读,而是在人文主义的观照下,对都市发展、人类命运做着奇思妙想般的考察。

"爱屋及乌"是诗,诗让栖居的意义在本质上得以实现。我们追求物质世界的富足,渴望精神家园的美好,至上境界,是"诗意地栖居"。学会诗意地生活,收获诗意的人生,需要我们有爱,而爱,是对天地的顺应与敬畏,是源自内心的安详与和谐。

"爱屋·及邬"作为展览,当然和邬达克有关,也和邬达克的建筑有关,但它不是一个怀旧的文献展,也不是炫技玩酷的建筑效果图陈列,而是一次直面当下、接续历史的讨论和对话。展览所邀请的艺术家,大都对都市或建筑有着长期的思考和积累,并以各自的实践成果享誉业界。他们将和你一起,相约邬达克,以回望的姿态,做一次"爱屋·及邬"的还乡之旅。

Love Home, Love Hudec

Painting and Sculpture Invitational Exhibition in Memory of L.E. Hudec

By Li Xiangyang

Sheltering in the depths of No. 129 Panyu Road is a Tudorbethan villa with clear excess and recess on the façades, whose interior decoration is embellished to perfection. The garden is nonexistent any longer and the horizon is blocked by surrounding skyscrapers. Fortunately, there is an eye-catching dovecote nestled on the west gable wall. Several motionless doves on the habitat are so lifelike as if cooing towards the sculptured birds on the roof, which are answering back. All this creates a cozy scene of home with love in balanced serenity and vitality.

Everyone loves his home. A bird cares about its nest. Compared to peoples from other civilizations, we Chinese are more reluctant to repair our house and mow the lawn in person; however, our expression of our love for our own home is all the more humorous and profound. A couple can "get divorced" and brothers can fall out unhesitatingly even without any regard of their children's fortune and future. Few Chinese do really care about their neighbors. Forced demolition takes place now and then with muck and mire flying all the way. In this circumstance, it is adequately a great contribution to the current society if you keep your house in good order and observe "Seven Don'ts" in public. So, "to love one's own home and its environs as well", to some extent, is a faith to be publicized. "To love one's own home" is just one man's nature and instinct while "to love its environs" is nurture and intelligence.

Certainly, it is the architect who ignites and materializes the love of home and endows it with life and beauty. Nothing can be compared to the architecture which adopts our lives from birth to death and accepts each of our expression of feelings, be it love or hatred. Different architects design different lives, which are embodied in various forms of home. Undoubtedly, Laszlo Hudec is an outstanding architect. His success does not only lie in the achievement in shaping Shanghai what is like now by providing Shanghai with more than 60 excellent historic buildings including hotels, residences, churches, hospitals and cinemas, nor simply in the legend of an individual and the architectural history of a city where he contributed exotic architecture and skillful construction technology, but also in his insight into clients, policies, environments and trends, his acute sense of different cultures, and his open and easygoing mind towards God, nature and fate as well. Today, we commemorate Laszlo Hudec because we cherish the form and the quality of buildings in the past. Today, we follow Laszlo Hudec because we cannot forget the impressive and sympathetic stories which took place under these familiar roofs .

The artists are another type of architects who create for us the spiritual home. They may have a perspective distinct from that of an architect, but they concern the state of our life as the architects, and inspire us to question the ultimate relationship between man and the world. Modern civilization has stereotyped the entire world, which disrupts our original lifestyle . Many cities have almost the same appearance and are getting more and more similar to one another. How will the people from concrete forests face their home? How will they hold fast to their independent spirit and individuality? It seems that there is no standard or reference for this. For many years, we run too fast physically in the fast process of powerful urbanization while our spirit falls far too behind. As a result, we live enthusiastically and vigorously without but feeling empty and lonely within. The architect shelters us from wind and rain. Similarly, the artists settle our lost souls with their incisive and passionate works which touch our innermost. They paint the architecture, but their concern is not certainly architectural esthetics. They paint the street, but the paintings are not for studying the texture of the city. They are researching the development of the city and the fate of humankind in light of humanism.

 "To love home and its environment" is poetry which embodies the essence of dwelling. We pursue a rich material world. We crave a good spiritual home. The ultimate end is that "Poetically Man Dwells". Love is required for us to learn to live poetically and to reap a poetic life. Love is obedience to and respect of nature, and love oozes from our inner peace and harmony.

The exhibition of "Love Home, Love Hudec" is, certainly, related to Laszlo Hudec and his architecture. However, this exhibition is neither a nostalgic display of old documents, nor a show-off of skillful architectural renderings, but the dialog between the past and the present, and the discussion of how to go on in the future. Most artists invited to this exhibition have considered a lot and accumulated experience in metropolis or architecture for a long time and whose works have won a good reputation in art circles. They will accompany you together with Laszlo Hudec to embark on a journey of "Love home and its environment" as if returning to your hometown.

结缘上海　追求摩登

PURSUIT OF MODERNITY IN SHANGHAI

结缘上海 追求摩登

"'爱屋·及邬'—纪念邬达克绘画雕塑邀请展"断想

徐海清

邬达克的人生具有浓郁的传奇色彩；其多元善变的设计理念、摩登风格与上海包容、开放的城市精神高度契合，上海是邬达克功成名就的福地；纪念邬达克的"爱屋·及邬"画展，其创作、观赏过程都需要想象力。

1914年6月28日，奥匈帝国王储斐迪南大公携储妃在帝国领地波斯尼亚的首府萨拉热窝巡视时遇刺身亡，史称"萨拉热窝事件"。这一事件成了二十世纪第一次世界大战的导火索，点燃这根导火索的行刺者是20岁的塞尔维亚族青年普林西普。战争改变了无数人的命运，其中之一是与普林西普差不多同龄的奥匈帝国臣民——那年以优秀成绩毕业于匈牙利皇家圣约瑟夫理工大学建筑系的拉斯洛·邬达克。邬达克不得不放弃当建筑师的理想，应征入伍，成为奥匈帝国的军人，开赴对俄罗斯作战的前线。据不完全统计，在一战中丧命战场的军人约有千万人之多，邬达克有幸未在其列，但他受伤沦为俄国的战俘——此事发生在一百年前的1916年。对交战国而言，实在只能算小事一桩；但对邬达克来说，却成了他此后一生中传奇故事的序幕。

大战给国际社会带来了巨大损失和沉重灾难；同时，一些国家在政治、经济、科技、文化以及军事等许多方面也获得了长足的发展；历史赐予战后国际秩序重建的机会，民族意识的形成、民族观念的勃发、民族国家的纷纷建立，是这场战争客观上带给人类的进步成果，这是战争发动者和撕杀双方都始料未及的。

福兮，祸兮！祸兮，福兮！世事如此，人生亦然。

被流放到西伯利亚战俘营、时年23岁的邬达克，他的人生记录在两年后翻开了他做梦都难以构思的篇章——回家的念想虽未能达成，但侥幸得以脱离死亡威胁。一番颠沛流离之后，竟然逃到中国境内，途经哈尔滨，于1918年10月26日抵达上海。半个月后，大战宣告结束，奥匈帝国解体。邬达克这个战败国的前尉官，囊中空空，踯躅在貌似战胜国的中国最大城市上海。不过，邬达克与近代上海开埠以后怀着淘金梦纷至沓来的外国"冒险家"不同，初来乍到的他恐怕并没有把这个"冒险家的乐园"当作久留之地；在他心中，这个城市只是一个驿站。一旦积攒了足够的盘缠，他将随时踏上归途。

然而，命运之神眷顾了这位出身于著名营造商家庭、青少年时期已获得木匠、泥水匠和石匠证书、尔后成为大学建筑系高才生的年轻逃亡者。真所谓"天无绝人之路"，上海的匈牙利救济会给逃亡者邬达克提供了落脚谋生的机会。同年11月初，他就在一家建筑行获得了一份工作，进而成为公司合伙人，顺利回归了建筑行业。随着战争的结束，邬达克凭借自己的天赋和努力，开始了他施展自己才华的岁月，不久便声誉鹊起，一颗明星在上海建筑界冉冉升起；不久，邬达克在上海娶妻成家、生儿育女。到了1930年代，他的事业更是如日中天。上海，这个相对邬达克的祖国所处的中欧地区十分遥远、相对他所属民族的文化十分隔膜的远东城市，成了他这个倒霉的逃亡者安身立命、功成名就的福地！

可是，同样令邬达克未曾预料到的是，正当他的事业蒸蒸日上之时，第二次世界大战爆发，匈牙利加入了德国、意大利、日本轴心国集团，而邬达克身不由己地被任命为匈牙利驻沪领事馆荣誉领事。进入1940年代后，时局使这位建筑奇才职业生涯的鼎盛期无可挽回地终结。1947年，作为再次成为战败国的子民和曾经的外交官，邬达克携家人秘密离开上海。他在游历欧洲各地后没有回到他经常魂牵梦萦的故乡，而是去了美国定居，这一选择不外乎是出于政治考量。但令人不解的是，在旅居美国的11年里，他没有再操建筑设计旧业、再创辉煌，而是潜心于考古研究。邬达克离开上海后不久，与他的故国一样，中国社会也发生了政权易手、天翻地覆的变化。由于各种原因，曾经书写了上海现代建筑史辉煌一页的邬达克的名字一度很少被人提及，而其行状则更是鲜为人知了。

在一段时期里，当人们谈及19世纪后期至20世纪上半叶上海西洋风建筑时，都会首推在外滩沿黄浦江1500多米地带上矗立的近30座具有各种时代特点、体现各类设计理念和风格的建筑群。它们被誉为"万国建筑博览"，美轮美奂，其景观地位无可替代。但是，这一时期的上海，西洋风建筑及西洋建筑群，实不止存在于外滩一带。由邬达克设计的近百栋风格迥异的楼堂馆它、剧院厂房，像一颗颗建筑艺术的珍珠，散布在外滩之外的上海各处，主要位于中西部的租界及其毗邻地区。

进入新世纪后，同济大学建筑与城市规划学院研究生华霞虹（现为同济大学建筑系副教授）在其导师——当今中国著名建筑学专家、中科院院士、法国建筑科学院院士郑时龄教授指导下完成的论文《邬达克在上海作品的评析》，开启了当代系统研究邬达克和邬达克建筑的先河。经过一个甲子岁月的洗刷，历史上邬达克的国籍背景和"政治面貌"，在更加开放、更趋多元的中国社会，尤其是在位居改革开放前沿的上海，已游离于人们在乎之外。十多年来，人们追寻着历史上的邬达克，思索着邬达克的成就的当下意义。他的旧居得到了有识之士的修复并被辟为纪念室，有关他的著述和视觉作品不断问世，在上海城市历史文化遗产保护制度化建设不断完善的背景下，"上海邬达克建筑遗产文化月"活动成了市民文化生活中的一个热点。

媒体人兼作家王唯铭曾写过两本关于上海近现代建筑史的文化读物，一本是出版于2007年的《墙·呼啸：1843年以来的上海建筑》，邬达克及其作品在该书的一节中大约占了两页篇幅；几年之后，王氏另一本关于上海建筑史的著述《与邬达克同时代：上海百年租界建筑解读》面世。该书的内容布局、章节标题与书名逻辑上的一致，使作者欲表达的观点跃然纸上：邬达克是上海近代以来建筑舞台上具有统领全局意义的主角、是这个时代建筑的符号，地位之高无以复加。也许有人对此持不同见解。不过，以下两句话应该是值得重视的：时任文新报业集团社长陈保平指出："邬达克之于上海，是20世纪城市文化风貌转折的一杆帆影。"华东建筑设计总院首席总设计师汪孝安认为："邬达克在上海建筑实践的二十多年则仿佛上海近代建筑发展的一个缩影。"

近些年来，不少人对邬达克能够成为"上海建筑界教父"的原因作了分析，比如：他以优质的作品树立信誉赢得了市场，他擅长因地制宜进行设计构想，他善于传播西方先进的建筑理论技术等。

据说邬达克脾气暴躁，以至遇事看不顺眼、一言不合就会连骂带打，这种做派不知是否与他的一段从军经历有关。但是，他在涉及设计事务时，与他的客户相处却表现得特别的耐心。据邬达克的女儿回忆说，她父亲不会把自己的设计方案强加于人，总是先认真倾听业主的想法，"然后再选择适当的技术将业主的想法建造出来"。这种给予各种客户自由选择的灵活性，可以说是他在设计能力上具备多元善变底蕴的展示，也不妨可看作他在上海这个各行各业激烈竞争的逐鹿之地深谙经营之道的表现。这种智慧同样表现在与营造商合作过程中，旧上海的几家大营造厂老板，与他关系都挺好。如馥记的陶桂林、洽兴（Yah Sing Construction Co.）的王才宏，都是上海乃至中国建筑业界的大佬，邬达克与他俩联手的最著名代表作是国际饭店和上海"颜料大王"吴同文的私宅"绿房子"等。邬达克入境随俗，广结善缘，朋友中有时任国立交通大学校长的孙中山先生之子孙科。孙科在邬达克资金周转困难时曾帮过他一把，于是他把本来准备自用的一套约一千平米的花园住宅低价转让给了孙科——投桃报李，中国人传统的人情世故他也了然于胸。邬达克开设的建筑事务所，30名雇员中一半是中国人。据说，他还学过沪语。这些，或许都应列为他的成功原因。

我以为，在邬达克方面，成功的根本原因或可概括为一句话——建筑设计理念的多元善变。关于这一点，已有不少业内人士作过解读，郑时龄先生的阐述无疑更为精湛全面："邬达克是上海新建筑的一位先锋，他善于学习世界各国的建筑式样，孜孜以求建筑的时代精神。他的建筑风格历经新古典主义、表现主义、装饰艺术派以及现代建筑风格，仿佛建筑风格的大全，既有当时欧美建筑的直接影响，也有建筑师个人的创造……上海培育了邬达克，而作为现代建筑的倡导者，他也创造了上海建筑的摩登风格。"

惟多元善变才能达到摩登境界。对邬达克的"摩登风格"，郑先生另有相关表述为："新潮设计"，"总能有新颖的构思，思路从未枯竭"，设计"没有重复出现的母题"。当一位建筑师这些十分难能可贵的素质落到了当年华洋杂处的上海滩、特别是租界及其毗邻区域，无异于如鱼得水。上海这座近代中国最早步入 process of modernization（现代化进程）的城市，天然是 modern style（摩登风格）的最佳生存、弘扬之地。邬达克的多元善变与上海的包容、开放，两者契合度之高，当年没有任何一座中国其他城市可以与之相颉颃。这是邬达克得以成为"上海建筑界教父"的客观原因。

遥想当年，这位建筑奇才手持一本山寨俄国护照，几经辗转，逃到上海。二十多年里，先以俄国公民的身份谋得第一份差使，然后他申领、变换过捷克斯洛伐克、匈牙利的护照，直至1940年正式获得匈牙利国籍并得到永久护照。而邬达克的出生地——当年属于奥匈帝国的匈牙利王国的泽尔伊欧姆省拜斯泰采巴尼亚、后来归属斯洛伐克；邬达克在美国故去十二年后魂归斯洛伐克，安葬在他幼年时受洗教堂边的家族墓地。有鉴于此，现在匈牙利和斯洛伐克两国都把邬达克当作自己的乡贤，分别举办了各种纪念活动。对此，中国方面表示了充分的理解和尊重，多次与两国合作开展与邬达克相关的文化交流。上海相关部门负责人给了邬达克一个很精准的"头衔"——"旅沪斯裔匈籍建筑师"——这般地域概念的糅合，颇能体现中国传统文化的中庸之道，即"合适""刚刚好"之意也。

历史上斯洛伐克和匈牙利关系极为密切，而作为一位杰出的建筑设计师，邬达克的事业在中国上海，上海是他的第二故乡。如果说，当年邬达克改变了上海的建筑风貌，那么，是上海这个中国近代以来经济文化相对发达、开放的沿海大城市、"大码头"，使他乐于，也有可能在一生中精力最为充沛的阶段，给上海留下了一张张建筑文化名片。"认识邬达克，也是认识上海城市的历史和未来"（郑时龄语）。无论在事业上还是个人生活上，邬达克都融入了上海，而上海则欣然接纳了这位来自异国的精英。基于邬达克与上海的因缘际会和他对上海的卓越贡献，从某种意义上说，称邬达克是"上海最著名的建筑设计师"或许与他的专业生涯和人生轨迹更加匹配。

邬达克的建筑艺术作品成了上海历史文化遗产的有形组成部分；而邬达克多元善变追求摩登的设计理念，已然顺理成章地积淀在海派文化之中。邬氏设计理念作为无形的精神遗产，对当下作为"设计之都"的上海而言，更显得弥足珍贵。我们把邬达克建筑作品列为历史文化遗产加以保护并开展各种纪念邬达克的文化活动，借助这一文化遗产进一步推进中外文化交流、交融，促使"上海设计"融入、引领世界潮流——这不妨可看作是新时期我们对"摩登"的诠释。

在邬达克沦为战俘整整一百年后的日子里，作为第二届"上海邬达克建筑遗产文化月"活动项目之一、以"爱屋·及邬"为主题的画展将在他的旧居展出，作品将汇集成册。

有介绍文字说，邬达克年轻时十分喜欢画画。不知道是否有邬达克的画作传世。尽管未曾见到他的绘画作品，但可以料想的是，作为建筑专业的学生，他习画的题材大抵不会缺少建筑物吧。

建筑与绘画同属视觉艺术，中外美术史都将建筑艺术的发展沿革作为其重要组成部分。建筑是技术与艺术的综合体，它属于三维造型视觉艺术形式；建筑设计师除了让建筑具有实用性功能外，还需赋予它们艺术审美的观赏功能。绘画是运用线条、色彩和形体等艺术语言，通过造型、设色和构图等手段，在二维空间里塑造出静态的平面视觉形象，以表达作者审美感受。需要指出的是，绘画虽然是在二维空间上塑造平面视觉形象，但因应透视原理，画家依靠明暗和形象结构表现物象的凹凸，造成立体幻象，并通过物象大小、光影明暗、遮挡关系、透视变化和色彩配置等手法，使绘画在二维平面上，呈现立体的空间效果。尽管画家在二维平面上所画的形象只能是现实世界三维空间中实际物体的一部分，但人脑意识对观察到的对象具有能动的反映功能，即能够通过部分将其想象成完整的状态。这就是画家能以二维平面再现三维立体空间的艺术奥秘之所在。

建筑师都会画画，大抵主要是画"建筑画"。据贾倍思、赵军编写《大师建筑画》所述，所谓建筑画是建筑师展示与完善其设计构思的表现手段之一；建筑画不仅是建筑效果图，建筑设计师从构思初期的草图、平面图、立面图到立体效果图都可以说是建筑图。从这些图中，人们可以领悟到建筑设计师对建筑形式与空间的整个探索与创造思维的过程。

这样一种定义的"建筑画"虽然也有某种艺术性，但与画家以建筑物为描绘、表现对象的绘画艺术显然是不同的。

英国学者理查德·豪厄尔斯（Richard Howells）在他的《视觉文化》（*VISUAL CULTURE*）一书中反复强调，绘画艺术是"想象生活的表达，不是实际生活的复制"。应该把绘画作品看作是对情感的表达。简而言之，画家要超越"所见即所得"的局限，这样的视觉文本才能奇迹般地创造出语言所不能表达的东西。

画家画建筑物通常都会凸出起某些部分，或辅之以其他物象"代码"，藉以强化作品的文化内涵和画家本人的思想情愫。

豪厄尔斯对观赏者则提出了审美路径上的要求：仅仅从表面上去欣赏视觉文本，我们将无法透过表面探索其深层的意义……艺术家期待观众能理解其（绘画）内在含义。然而，这些含义的表达依赖于画家与观众之间共同的文化传统。了解那些"代码"。换而言之，观赏者要"透过表面探索"作品"深层的意义"，同样需要借助想象力。

笔者孤陋寡闻，在我的印象中，西方绘画史上，为人们熟知并以一睹原作为快、津津乐道的作品基本上是历史、宗教题材，是人物画，风俗、风景画，以及静物。以建筑物为主体或含建筑物元素的传世名画似乎并不多。但在我看来，即便建筑物在绘画中只是作为背景或气氛的烘托，都是画面上不可或缺的有机组成部分，有的更是起了很关键的作用。文艺复兴时期意大利画家拉斐尔（Raffaello Samzio 1483——1520）的《雅典学院》（*The School of Athens*）和俄国巡回展览画派（Peredvizhniki）画家苏里科夫（Vasili Ivanovich Surikov）的《近卫军临刑的早晨》（*The Morning of the Streltsy Execution*），都堪称与建筑物相关的最为经典的名作。

大型壁画《雅典学院》以古希腊哲学家柏拉图创建的雅典学院为题材，表彰人类对智慧和真理的追求。大厅里汇集着不同时代，不同地域但都以古代七种自由艺术——即语法、修辞、逻辑、数学、几何、音乐、天文为研究对象的不同学派的著名学者，有以往的大家，也有当代的名人，他们正在激烈地辩论。画面上人物众多，姿态各异；建筑物与人物堪称平分秋色——作品以纵深展开的学院的拱门为背景，乳黄色的大理石结构与人物衣饰的各种颜色非常协调，而拱门的高大则象征着这些人物对希腊精神的崇拜和对最高生活理想的追求。作品对建筑物的成功处理与拉斐尔是画家与建筑师一身而兼二任不无关系吧。

创作于1879年、以1698年俄国沙皇彼得大帝镇压近卫军兵变这一事件为题材的油画《近卫军临刑的早晨》，选取了兵变失败的近卫军在莫斯科红场临刑前的悲壮时刻构建画面，极为震撼人心。画面的背景是莫斯科克里姆林宫的外墙和瓦西里·伯拉仁诺大教堂，教堂大尖顶和9个洋葱头状的圆球顶，十分壮观。教堂建于1550年代中期，是俄罗斯历史上第一位沙皇伊凡雷帝为纪念战争胜利而建造的。据说，为了不在别处再出现这样美丽雄伟的教堂，伊凡雷帝居然下令弄瞎了设计这座教堂的建筑师的双眼！这些高大而富有历史内涵的建筑物，有效地衬托了刑场肃杀气氛，彰显了彼得大帝的强势地位。

而16世纪尼德兰地区最伟大的画家彼得·勃鲁盖尔（Pieter Bruegel，约1525——1569）创作的以圣经故事为题材的《巴别塔》(*Babel*，一译《巴比伦塔》、《通天塔》)，则是西方绘画史上少有的由建筑物几乎充满画面的杰作。

传说在大洪水以后，诺亚率众人来到示拿这地方从事生产、繁衍后代，当地人口多了起来。他们担心再遭洪灾，决定修造一座高塔，以备水患时藏身。这座塔要容纳所有的人，就须高到能通达天庭。上帝耶和华得知此事，亲临人间看个究竟，当他见到正在建造的高塔后，十分惊恐，心想人类能造出这样的高塔，还有什么事做不成？为了阻止通天塔的建成，他施神术让示拿人说各种语言。正在建塔的示拿人因而彼此不能再顺利沟通，工程无法进行下去并且发生许多纠纷。结果塔未建成，人类只好按语言能相通为群散居各地。这座建了一半的塔，被称为巴别塔（"巴别"在古希伯来语中为变乱之意）。作品寓意深刻，但人们对它的理解并不相同，各有各的想象。有说它表现了人类在改造世界时与天意的不可调和性，揭示了人类为追求新生活而面临的悲剧，也有说它影射了现实世界的混乱与纷争。

巴别塔是虚幻的建筑，勃鲁盖尔用的是写实手法。他画了云彩围绕塔顶，使之隐约可见，以此表现塔身之高。除了高塔，画面上有众多处于混乱状态的人物，画家拉开人物形象与塔身、大自然的比例和距离，以显示工程的宏伟与建造之艰巨，突出了人类的创造性力量。

与西方绘画史上建筑物在作品中权重偏小的情况不同，近二三十年来，建筑物入画在我国美术界并不罕见且佳作频出。应邀为本次"爱屋·及邬"主题画展进行创作的本市画家，都有相当高的艺术造诣，对通过建筑物的描绘表现现实生活和历史遗迹方面都有成功的实践。

他们大多属"中生代",都已成果斐然。

如杜海军先生以城市楼群为其主要创造题材,《N个窗》《城市精神》《城市表情》等都广受好评。杜海军的城市楼房画,常常是透过画面上各式楼房的一个个窗口,反映了人们的生存状态,激发观者由此对城里人心境的想象,含蓄地表现了画家对现代化、城市化的思考,他所画的是他"心中的城市"。

应海海先生多年来怀着一份"在这些饱含历史的老建筑里寻觅过去的风云变幻"的心情来描绘老建筑。他创作的石库门油画,构图精致,尤擅光影处理,使老房子充满了温馨,被誉为有"牧歌之美"。

洪健先生近些年来以都市建筑为题材的系列作品令人瞩目,他的《洋务遗存——上海百年水厂》让人一见之下便产生难以挥去的沧桑感。2009年,此画先后摘得首届上海"白玉兰美术奖"一等奖和首届中国美术奖创作银奖,确是实至名归。

曾有论者说毛冬华女士在"海上画坛堪称独门"——指她以水墨画的柔软表现砖木、水泥乃至钢结构、玻璃幕墙的坚固挺拔,效果别有一功。她的《乍浦路桥》(桥后是天主教新天安堂旧址)、《多云转晴》(画的是金茂大厦),是同类题材中的精品。近年来,毛女士在建筑题材绘画的探索中,更加侧重"研究探索绘画本体语言的丰富性及厚重感"。

同为出生于上海的陈健先生和李乾煜先生,与上述几位中青年画家一样,他们的作品都多次获得过全国、上海市的各类奖项,不少佳作被美术机构和重要公共场所收藏。他们几乎都在艺术院校执教,有的还是硕士生导师。

与"中生代"相比,年岁稍长的是李向阳先生、张安朴先生与贺寿昌先生。

本次画展的策展人李向阳先生,在阐述"爱屋·及邬"这一主题时说:"现代文明将整个世界格式化之后,我们原有的生活方式也被打乱,千城一面,日益趋同,那些从一样的混凝土森林中走出来的人,何以面对家园,何以坚守独立的精神和品格……许多年来,在迅猛的都市化进程中,我们的肉身走得太快,灵魂却落在了后面……。"他指出,艺术家"画建筑,不一定关注建筑美学的讨论,画街景,也不是对城市肌理的研读,而是在人文主义的观照下,对都市发展、人类命运做着奇思妙想般的考察"。这些话语,既有哲学、社会学的深邃涵义,又不乏浓浓诗意,不愧为曾在美术院校、团体担任过重要职务的资深美术家。

张安朴先生是地道的"老上海",曾任解放日报摄影美术部主任。他的硬笔水彩画作品中有上海城隍庙九曲桥、湖心亭和南京路步行街两边的楼房、古罗马遗址和巴黎圣母院、婺源老宅和香港孙中山纪念馆……这些作品以洗练的笔触刻画的不仅是建筑物形制,更反映出它们所在处中外古今不同韵味的文化氛围。2013年,中国邮政发行了张安朴设计的一套四枚特种邮票《豫园》,无疑是对他在建筑题材绘画创作领域的高度肯定。

二十世纪七十年代后期至八十年代,贺寿昌先生的作品曾多次获文化部和上海市的优秀奖项。然而,此后大约二十多年里,他因公务繁忙不得不搁下画业。近几年,贺先生终于握起久违的笔,画兴大发。他的作品中有好几个建筑题材系列,其中组画"城市与大师",将四位名画家——吴昌硕、徐悲鸿、刘海粟、丰子恺在上海的故居,置于春夏秋冬四个季节,用不同的主色调再现了四处不同的建筑,足以让人见屋思人。值得一提的是,他创作"城市与大师"的动机还在于:"当我们把这些大师作为上海的标志介绍给世界时,世界也会向上海致敬。"

诚如李向阳先生所言,"展览所邀请的艺术家,大都对都市或建筑有着长期的思考和积累,并以各自的实践成果享誉业界"。笔者对这些画家的简要介绍和评述,只是作为观画人的点滴感受,或是在对某些业内人士评论认同基础上的转述。篇幅有限,难免挂一漏万,顾此失彼;若有安议,还望画家和画评家们见谅。

我们完全有理由相信,应邀参与"爱屋·及邬"主题画展创作的画家们,都能通过与邬达克之间穿越时空的"对话",出色地在二维平面上再现由邬达克设计的同属视觉文化的三维艺术作品,或通过艺术地还原邬达克设计师职业生涯中的某个侧面,表现他与建筑心物一体的情怀。"爱屋·及邬"画展呈现的是艺术家对描摹刻画对象的主观感受,而观赏者同样需要展开想象的翅膀,才能充分见识到与邬达克建筑作品实体和对邬达克文字描述不尽相同的风貌。

2016上海书展期间,供职上海市作家协会的老同事吴越女士,在微信朋友圈晒出了华夏虹、乔争月所著《上海邬达克建筑地图》一书的照片,书封加内页共6幅,附有一句颇为感慨的话:"难怪我对邬达克莫名喜欢。"我问她,去过邬达克纪念馆吗?回答说,没有,也不知在哪里。我当然乐意陪她参观,不惟如此,还请她为这次纪念月活动出谋划策。而她说:"我要读书,好好了解一下这个人。"——这就是曾经担任过文汇报首席记者的文化人的品性。

与此同时,我又想,这几年来各方面已为介绍邬达克做了不少颇有成效的工作,但覆盖面还尚需不断扩展,深入度还有待继续发掘——这,不也正是本届"上海邬达克建筑遗产文化月"活动举办"爱屋·及邬"主题画展的初衷吗?

Pursuit of Modernity in Shanghai

My Impression of
" 'Love Home, Love Hudec' — A Painting and Sculpture Invitational Exhibition in Memory of L. E. Hudec"

By Xu Haiqing

L.E.Hudec's life is very much legendary. His diversity and variability in design and his modern style match well with the spirit of Shanghai - tolerance and openness. Shanghai is the city where Hudec was blessed with successes and achievements. "Love Home, Love Hudec" - a painting and sculpture exhibition in memory of L.E.Hudec demands imagination in creation and appreciation.

On June 28th, 1914, Archduke Franz Ferdinand, heir presumptive to the Austro-Hungarian throne and his wife Sophie, Duchess of Hohenberg were shot dead while inspecting Sarajevo, the capital of Bosnia, the domain of his Empire. This is called "Sarajevo Assassination" which triggered the outbreak of WWI last century. The assassin was Gavrilo Princip, a 20-year-old Serbian. WWI changed the fate of countless people, among which one is Laszlo Hudec, a subject of Austro-Hungarian Empire who was at nearly the same age of Gavrilo Princip. In that year, Laszlo Hudec just graduated with all straight A's from the Department of Architecture, Hungarian Royal Joseph Technical University. But alas, he had to give up his dream of being an architect, and was enlisted in the military of Austro-Hungarian Empire to fight against Russia. According to incomplete statistics, about ten million soldiers were killed in WWI. Fortunately, Laszlo Hudec survived the war. He was injured and captured by Russians in 1916, which is a small issue for the two warring states about a century ago, which marks the beginning of the legendary life of Laszlo Hudec.

Although WWI caused great damage and disaster to the international community, yet meanwhile some nations achieved sufficient development in politics, economy, technology, culture and military. History gave an opportunity to establish a new post-WWI international order. The formation of national consciousness, the thrife of nation sense and the successive establishment of nation-states were, objectively human progress brought about by WWI, quite beyond the presumption of its instigators and the warring opponents.

Is it a blessing or misfortune? Or is it that misfortune turns into a blessing? Who can tell? Everything is hard to define. It is the same for life.

Laszlo Hudec aged 23 was exiled to the Siberian Concentration Camp. His life opened a new chapter that he had never ever dreamed of. Although he was unable to return to his homeland he missed badly, yet he fortunately escaped from the threat of death. After being homeless for a long while and tramping, he fled to China. He passed by Harbin and arrived in Shanghai on Oct. 26th, 1918. Half a month later, WWI was over, and the Austro-Hungarian Empire was disintegrated. Hudec, an ex-officer of the defeated country, with nothing possessed, wandered in Shanghai, the biggest city of China, a so-called victor of WWI. But different from other foreign "adventurers" pouring to Shanghai in search of a fortune after Shanghai became a port in the 19th century, Hudec, a newcomer, very possibly didn't take Shanghai, "the wonderland for adventurers" as his permanent residence. Shanghai was just a posthouse. When he accumulated enough money for fare, he would go back home at any time.

Hudec came from a famous family of constructors. He had got a carpenter's certificate, a plasterer's certificate and a mason's certificate at a young age and become a gifted college student in the Department of Architecture. And now the war turned him into a runaway. However, Goddess of Fate favored Hudec. You see, there is always a way out. The Hungarian Relief Association in Shanghai gave him an opportunity to make a living. At the beginning of November 1918, he got a job in an architectural design firm and then became a partner of this firm. He returned to architectural industry successfully. WWI was over and Hudec started his career with his talent and strife. Before Long, he became well known as a bright star rising up in the circle of Shanghai architecture. Shortly after, he got married and had children in Shanghai. In the 1930s, his career came to the climax. Shanghai, an oriental city which is far away from his homeland in Middle Europe and has a great cultural gap for him, became a promised land for Hudec, an unfortunate fugitive to settle down and to make great achievements.

But when his business was on the rise, another unexpected event shocked him. World War Two broke out. Hungary joined Axis Powers allied by Germany, Italy and Japan. Hudec was appointed as the honorary consul of Hungary in Shanghai, which was against his will. In the 1940s, the situation made a fatal end to his business in his prime time. In 1947, as a subject and a diplomat of the country defeated again, Laszlo secretly left Shanghai with his family. He travelled across various countries in Europe, but didn't return to his homeland that he was longing for so desperately. Unexpectedly, he settled down in the USA. This choice is perhaps much of political concern. But it is beyond our expectations that he had never been involved in architecture for 11 years in the USA. Instead, he was buried in his research into archeology. Shortly after he left Shanghai, China experienced the change of political powers similar to that of his homeland. Laszlo Hudec, who had contributed a glorious page to the history of Shanghai modern architecture, was seldom mentioned for many reasons, and his story was barely known to anyone.

For quite a long period of time, when people talk of western-style buildings during the later 19th century and the first half of 20th century, they prefer the clusters consisting of nearly 30 buildings with design concepts and styles of different ages, standing along a belt of more than 1500 meters on the Bund along the Huangpu River, which is known as "Exotic Building Clusters". They are beautiful and irreplaceable scenic spots. But in the same period, there were many other western-style buildings and clusters in Shanghai. Hudec designed nearly 100 buildings with quite different styles including mansions, residences, theatres and factories, which are like pearls with architectural glamor, dispersed in various corners of Shanghai beyond the Bund, mainly in the central and western concessions and their neighboring areas.

In the 21st century, Hua Xiahong, a postgraduate of the School of Architecture and Urban Planning, Tongji University (now Associate Professor of the Department of Architecture, Tongji University) finished her dissertation "Analyses on Hudec's Architecture in Shanghai" under the instructions of Professor Zheng Shiling(a renowned Chinese expert of architecture, a member of Chinese Academy of Sciences and a member of French Academy of Architectural Sciences) which pioneered the systematic research of Hudec's life and his architecture. Six decades later, Hudec's nationality and "political status" are no longer concerned by the people in a much more tolerant and diverse society of China, especially Shanghai, a pioneer of reform and opening to outside world. Over a decade, we have been searching for information about Hudec in the long river of history and thinking about the significance of his achievements to the present society. His residence has been renovated by a respectful person and has been opened as Hudec's Memorial Hall. Research works and visual art works on Laszlo Hudec mushroom to the public. With the constant improvement of Shanghai urban historic and cultural heritage protection systemization, "Shanghai Hudec Architectural Heritage Cultural Month" event comes out as a hot spot in urban citizens' cultural life.

Wang Weiming, a special correspondent, editor and writer wrote two popular books on the history of Shanghai modern architecture. The first book is "Wall • Wuthering: Shanghai Buildings after 1843" published in 2007, in which the introduction of Hudec and his works take about two pages. Several years later, his second book on the history of Shanghai architecture-"Contemporaries of Laszlo Hudec: A Century's Foreign Concession Architecture in Shanghai" went to press. Its contents and chapter themes match with the logic of book name so well that you can immediately get his opinion across the lines: Hudec plays the leading role on the stage of Shanghai modern architecture and is symbol of architecture in that period, and his achievements are absolutely insurmountable. Someone may have different ideas. But the following words shall be agreed upon unanimously: "Hudec is a mast of the change of urban cultural outlook for Shanghai in the 20th Century," as pointed out by Wu Baoping, once the president of Wenhui Xinnin United Press Group. "Hudec's 20 years' practice in architecture is like a miniature of the development of modern architecture in Shanghai", remarked Wang Xiao'an, Chief General Designer of East China Architectural Design

Institute.

In recent years, the reasons why Hudec can be honored as "Godfather of Shanghai Architecture" have been analyzed and summarized as follows: "He won the market with his quality works", "He is good at planning according to the local conditions", "He is good at spreading the western advanced architectural theory and technology", etc.

It is said that Hudec had a fiery temper. He would curse and beat someone if there was something he didn't like about him or a word that irritated him. This behavior is perhaps related to his military experience. However, when he was negotiating design business with his clients, he would be excessively patient. As his daughter recalled, her father never imposed his design plan on others, and he would listen to the client's ideas attentively, "and then he would choose a proper method to materialize the client's ideas." This flexible range of options for the clients is the demonstration of both his diversity and variability in design and his principle of business in the ever-competitive industries of Shanghai. His wisdom was also well-demonstrated in his cooperation with local constructors. He had good relations with many well-known constructors. For example, Tao Guilin from Fu Kee Group and Wang Caihong from Yah Sing Construction Co. were great leaders in architectural industry in Shanghai if not in the entire China. Together with both, Hudec worked out many famous masterpieces such as Park Hotel and the great pigment merchant D.V. Woo's residence (nicknamed as "green house") etc. When in China Hudec did as the Chinese did. Sociable as he was, Hudec made friends with many VIPs, among which the most prestigious one was Sun Ke, the only son of Sun Yat-sen (the Founder of Kuomintang), president of Jiao Tung University. Sun Ke helped Hudec when he was short of fund. In return for Sun Ke's favor, Hudec transferred at a very low price the garden residence of approximately one thousand square meters (which he had planned to live in) to Sun Ke. This story is a good piece of evidence showing that he was very familiar with the traditional interpersonal relationships among the Chinese. In Hudec's own firm with 30 employees, 15 were Chinese. It is also said that he even learned to speak Shanghainese. All these can be listed as the factors of his success.

I think, the root cause of his great success can be summarized as diversity and variability in design concepts. Many insiders have made analyses of this point. Mr. Zheng Shiling's comment is incisive and complete: "Hudec is a pioneer of modern architecture in Shanghai. He is good at learning architectural styles from various countries in the world. He was devoted to his zealous pursuit of the zeitgeist of architecture. His styles varied from Neo-Classicism, Expressionism, Art Deco all the way to Modern Architecture. He is like an encyclopedia of architectural styles, a combination of direct influence of European and American architecture with his own creativity……Shanghai bred Hudec. In return, Hudec, an advocate of modern architecture, modernized the architecture for Shanghai. " Diversity and variability is the sole access to modernity. Mr. Zheng described Hudec's "modern style" as a "trendy design", "he always worked out novel solutions, and his inspirations never stopped", and his designs "have no repetitive motifs". When Hudec, an excellent architect with such precious qualities, set foot in Shanghai, a city blended with Chinese and westerners, especially the foreign concessions and their neighboring areas, he was like a fish swimming freely in the river. Shanghai is one of the earliest cities undergoing modernization. Naturally, it is the best place for modern style to survive and thrive. Hudec's diversity and variability matches perfectly with the tolerance and openness of Shanghai that no other city in China could claim to possess. This is the objective reason why Hudec can be "Godfather of Shanghai Architecture".

Imagine: this genius in architecture with a pirated Russian passport, encountered many obstacles and finally fled to Shanghai.,. In the coming twenty years, he took the first job as a Russian citizen, and then applied for a Czechoslovakian passport and changed it to a Hungarian passport. Finally, in 1940, he formally applied for Hungarian citizenship and obtained a permanent passport. Meanwhile, Besztercebánya once a part of Hungarian Kingdom, Austro-Hungarian Empire, the birthplace of Hudec, later became the soil of Slovakia; 12 years later after he died in USA, Hudec's ashes were taken back to his homeland, Slovakia and buried in his family graveyard near the church where he got baptized as an infant. Thereby, both Hungary and Slovakia take Hudec as their native honorable squire and hold many commemorative events respectively. All this is understandable to and respected by our Chinese government that's cooperated with both countries to hold cultural exchange events related to Hudec. The one in charge of the department of Shanghai Municipal Government has given Hudec a precise "title"- "Slovakian Hungarian Migrant Architect in Shanghai", a title integrating his race, nationality and immigrant residence into one, which embodies "Zhongyong zhi dao"(the medium) in traditional Chinese culture, meaning "suitable", "just fine".

Throughout history, Slovakia and Hungary had very close ties. Hudec, an outstanding architect from both countries, had his career in Shanghai--his second hometown. If it is acceptable to say that Hudec changed the architectural landscape of Shanghai, then it is equally acceptable to say that it is Shanghai, an open coastal city and "the great dock" with more advanced economy and culture in modern China that delighted him and inspired him in his heyday to present many architectural and cultural cards of Shanghai. "To know Hudec is to know the past and the future of Shanghai." (As Zheng Shiling has claimed) Hudec got accustomed to Shanghai in both business and private life. Shanghai was also very glad to admit him, a foreign genius. Based on the perfect match of Hudec with Shanghai and his excellent contributions to Shanghai, in a sense, it is probably more suitable to say Hudec is "the most famous architect in Shanghai" as far as his career and life course are concerned.

Hudec's architecture and artworks have become a tangible part of Shanghai historic and cultural heritage. His diversity and variability and pursuit of modernity in design concept, has naturally enriched Shanghainese Culture. Nowadays, Hudec's design concept as intangible spiritual heritage is far more precious to Shanghai as "the Capital of Design". We list his architectural works as historic and cultural heritage under protection and hold various cultural events to commemorate Hudec, and take the cultural heritage, his former residence, to enhance cultural exchange and interaction between China and other nations so that "Shanghai Design" may become a trend setter in the world — which can be our new interpretations of his "modernity" at present.

One hunderd years later after his imprisonment in Russia as a POW, a painting exhibition is held on the theme of "Love Home, Love Hudec" in his former residence as a program of the Second "Shanghai Hudec Architectural Heritage Cultural Month". And the exhibited paintings will be compiled into a book.

The introduction told us that when young Hudec liked painting very much. I wondered whether there are any paintings of his handed down to us. Although I never saw his paintings, I assume that he should have practiced painting buildings for he majored in architecture.

Architecture and painting are both visual arts. The development and evolution of the art of architecture is an essential part of the history of Chinese and western fine art. Architecture is a combination of technology and art. It is a form of three-dimensional visual art. The architect considers not only the practical functions but the aesthetic appeals as well. Similarly, painting is the static planar visual image formed by lines, colors, shapes and other artistic elements in a two-dimensional space to show the esthetic feelings of the artist. It shall be pointed out that although painting is a 2 D visual image, the painter himself applies the perspective principle to create a steric illusion with demonstrated convex and concave of the object based on its light and shade and shape, and express the three-dimensional effect on a two-dimensional space with contrastive sizes, shadings, perspective variations, coloration and other methods. Even if the image painted on the two-dimensional paper is only a part of the object in the actual three-dimensional world, the brain of the observer still has a proactive function to reflect the object as a whole. That is to say, his brain is able to imagine the whole shape by just seeing a part. This is the reason why the painter can represent the three-dimensional image in a two-dimensional space.

All architects paint mostly about architectures. According to "Masters Architecture Paintings" compiled by Jia Beisi and Zhao Jun, what we call architecture drawing is just an expressional method for the architect to visualize and improve his design planning. Architecture painting is not just architectural sketching. A sketch of the initial planning, a plane, an elevation and a steric sketching can be all included in the architectural paintings. From these drawings, the readers can get familiar with the whole process of searching and creating of the architectural form and space by the architect.

Such a definition of "architecture painting" is somewhat artistic, but obviously different from the the art that takes the buildings as the objects of representation.

Richard Howells, the British scholar stresses repeatedly in his *VISUAL CULTURE* that painting is "the expression of imaginary life, not the copy of real life". So, we shall take paintings as the expression of the artist's feelings. In one word, the painter shall go beyond the limit of "painting what he sees". Thus, his visual text can create miraculously what our vocal language cannot express. When the painter is painting a building, he will highlight some sections with other image "codes" to strengthen the picture's cultural connotations and his own thoughts and feelings as well.

Howells raises a requirement on aesthetic appreciation for the observer: if we observe the visual text only superficially, we cannot probe its deep meanings hidden beneath the surface….The artist expects the audience to understand it (the painting) profoundly. However, the expression of these meanings depends on the cultural tradition shared by the painter and the audience to know those "codes". In other words, the observer also needs imagination to get "deep meanings" by seeing "through the surface".

I am not well-read. .but, I have a vague impression that in the history of western painting, the most well-known, popular and enjoyable works are paintings of human figures, landscapes and customs mostly on the theme of history and religion. There are not many famous paintings with buildings as the main theme or with architectural elements. In my opinion, even if the buildings are just the background or for the atmosphere, they are still an integral part of the painting or even the dominant part. For instance, In the Renaissance period, The School of Athens by the Italian painter Raffaello Samzio and The Morning of the Streltsy Execution by Vasili Ivanovich Surikov, a Russian painter from Peredvizhniki make the most classic masterpieces related to buildings.

The big fresco The School of Athens takes the Academy established by Plato, the Greek Philosopher as its theme to honor human's eternal pursuit of wisdom and truth. The grand hall assembles famous scholars, some are dead, some are Plato's contemporaries–all great figures from various regions and from different schools of thoughts but all studying the seven free arts –grammar, rhetorics, logic, mathmatics, geometry, music and astronomy. They are debating vigorously. In the fresco, there are many figures with different postures. The architecture and the figures are equally highlighted in the fresco—the arch of the Academy is the background, the creamy marble structure matches very well with various colors of the clothing, and the grand arch symbolizes those great figures' admiration of Greek spirit and their arduous pursuit of the highest ideal of life. The successful handling of the architecture in the painting cannot be disengaged from the two roles of Raffaello Samzio –painter and architect.

Painted in 1879, The Morning of the Streltsy Execution , takes the solemn moment of executing the Guards in Red Square to demonstrate the event that Russian Czar Peter the Great crushed the rebellion of His Guards in 1698. The background is the wall of Moscow Kremlin and Saint Basil's Cathedral. The tall tent-roofed tower and the 9 onion-shaped domes are very spectacular. Saint Basil's Cathedral was built in the middle of the 1550s, to commemorate the victory of the first Czar of Russia, Ivan the Terrible . It is said that Ivan the Terrible even gave an order to blind the eyes of the architects who had designed the cathedral so that such a beautiful and magnificent cathedral would not appear in any other places! These grand and historic buildings set off the solemnity of the execution ground and show the powerful dominance of Peter the Great.

In the 16th century, Pieter Bruegel the elder, the greatest painter in Netherland, painted a masterpiece based on a story from the Bible – The Tower of Babel, in which the Tower occupies nearly all the picture. This is seldom seen in the history of western painting.

As the biblical legend goes, after the great flood, Noah led his people to Shinar where they settled down and reproduced. Shinar became densely populated. The people were worried that there would be another deluge, so they decided to construct a high tower to shelter them from another possible flood. To hold all the people, the tower shall be tall enough to reach the heaven. Jehova, the God knew it, and came down to the earth to look at the tower that the people were constructing. He was shocked and terrified at the very thought that if the people can construct such a high tower, whatever else can they not do? To prevent the completion of Babel, He confounded their speech. Thus they could not understand each other, and the construction could not continue due to the communicative failure and lots of disputes that arose. Consequently, the tower was never completed, and the people scattered themselves around the world in groups that speak the same language. The uncompleted tower is called Tower of Babel ("Babel" means confusion in Old Hebrew). The painting has an incisive meaning. Its interpretation varies from viewer to viewer who imposes his own imagination on the painting. Someone says that it expresses the uncompromising contradiction between man's will to reform the world and God's will to keep its original state, which is an unveiled tragedy that man encounters in his pursuit of a new life. But someone else would say that it alludes to the riot and disputes of the real world.

The Tower of Babel is a phantom building. But Pieter Bruegel applied realistic method to the painting of it. He added clouds around the indistinguishable top to show that the tower is towering into the sky. Apart from the high tower, there are many people in chaos in the painting. The painter contrasted the proportion and distance between man, the tower and nature to show the magnificence of the tower and the difficulty in construction and thus the great power of man's creation is highlighted.

Architectural paintings only take a smaller portion in the history of western painting. In contrast, in recent twenty or thirty years, architectural paintings are not unusual in China art circles and many masterpieces are produced. All the painters being invited to present their paintings for "Love Home, Love Hudec" are very skilled, and very much practiced in expressing real life and historic traces by depicting the buildings.

Most of the artists are belonging to the "Middle Generation" who has already made great achievements. Here are several representatives.

Mr. Du Haijun mainly paints the city clusters. His N Windows, Spirit of the City, and Facial Expression of the City are widely popular and well received. Du Haijun's architectural paintings usually reflect the living state of the people by focusing on each window in different buildings, inspiring the observers to imagine the mental state of the urban residents and thus implicitly expresses his thoughts on modernization and urbanization. He paints "the city in his heart".

Mr. Ying Haihai has depicted the time-honored buildings for many years "in search of the turbulent past from the buildings full of stories and the sentimental memories of love as well". His Shikumen paintings are well structured. He is good at handling light and shade so that the old houses in his paintings are oozing warmth and love. His paintings are reputed as having "the beauty of pastoral peace".

In recent years Mr. Hong Jian has painted a series of urban buildings which have won him remarkable honor. His "Self-Strengthening Movement Heritage—Shanghai One Century Old Water Plant shows an unforgettable sense of history at first sight. In 2009, this painting won the First Prize of First Shanghai "Magnolia Award for Fine Art" and Silver Medal in First China Fine Art Award Contest. He deserves all the honors and reputations.

According to some critic, Lady Mao Donghua is a unique master among Shanghai painters", pointing out that she can successfully depict the solidity and straightness of wood-brick buildings, concrete buildings and even steel structure and glass curtain wall with the softness of traditional Chinese ink painting. Her *Zhapu Road Bridge* (with the former Catholic Union Church in the background) and Turning Clear (dominated by Jinmao Tower) are two masterpieces. In recent years, she focuses more on the expression of "the richness and severity of painting elements" in her exploration of painting buildings.

Mr. Chen Jian and Mr. Li Qianyu, both born in Shanghai, are the same as the above middle-aged painters. Their works have won many prizes at Shanghai level or national level. Many of their masterpieces have been collected by fine art institutes and other important public institutes. They are also teaching in art schools. Li Qianyu is also a master's tutor.
Compared to the "Middle Generation" painters, Mr. Li Xiangyang, Mr. Zhang Anpu and Mr. He Changshou are older.

Mr. Li Xiangyang, the curator of this exhibition, explained the theme of "Love Home, Love Hudec" –"Modern civilization has stereotyped the entire world, which disrupts our original lifestyle. Many cities almost have the same appearance and are getting more and more similar to one another. How will the people from concrete forests face their home? How will they hold fast to their independent spirit and individuality? …. For many years, we run too fast physically in the surging process of urbanization and our soul falls far far behind…." He added that "when artists paint the architecture, they are not necessarily concerned about architectural esthetics. They paint the street, but the paintings are not for studying the infrastructure of the city. They are researching the development of the city and the fate of humankind in light of humanism." These words are pregnant with philosophical and sociological meanings and also very poetic. He deserves the title of senior artist who has worked in the college of art and institutes.

Mr. Zhang Anpu is a real "Shanghainese". He was once the director of Photography and Fine Art Department of Jiefang Daily. His hard pen water colors cover Zigzag Bridge and the Isle Pavillion in Old City Temple, and buildings on both sides of the Nanjing Road Commercial Walking Street, Roman Relics, Notre Dame de Paris, Wuyuan Former Residences, and Hong Kong Sun Yet-sun Memorial Hall….These works are depicted with concise lines, to represent not just the shape of the buildings, but the cultural atmosphere with specific regional and historic flavors. In 2013, China Post issued a set of four postage stamps designed by Zhang Anpu, which is obviously a well-deserved recognition of his achievement in architectural painting.

From the later 1970s to the 1980s, the works of Mr. He Changshou had won many prizes from Ministry of Culture and Shanghai Municipal Government. However, in the following twenty years, he stopped painting because of his full schedule as a public servant. In recent years, he took up the long-abandoned brushes to paint with greater enthusiasm. He painted several series of buildings: one series is "City and Masters", putting the four residences of Wu Changshuo, Xu Beihong, Liu Haisu and Feng Zikai in Shanghai in four seasons- spring, summer, autumn and winter, and using four dominant colors respectively to present those buildings, which remind us of the masters who used to live there. What's worthy of mention is his motivation of "City and Masters" : "When we introduce our masters to the world as the name cards of Shanghai, the world will honor Shanghai."

Just as Mr. Li Xiangyang has said, "Most artists invited to this exhibition are those who have thought a lot and accumulated experience on metropolis or architecture for a long time and whose works have won a good reputation in art circles." I, as an observer, have little personal contact with them, and my introductions of them and related comments are based on my personal impressions of their paintings or quoted from some insiders whose views I agree with. For the limited space, I may have inevitably missed some important information. Please forgive me for any possible inappropriate judgments about any of those paintings.

We have every reason to believe that all the painters invited to work for "Love Home, Love Hudec" painting exhibition will "dialogue" with Hudec transcending time and space, and that they will be able to represent excellently the three-dimensional artworks designed by Hudec in two-dimensional planes or reproduce one side of Hudec's career artistically to show the spiritual unity of the architect and his architecture. What the exhibition demonstrates is the subjective feelings of the painters towards the objects. However, the observers need to take wings of imagination so as to see the difference between Hudec's architecture and others' verbal description of his architecture.

During 2016 Shanghai Book Fair, Lady Wu Yue, my former colleague with Shanghai Writers Association, shared 6 photos, including the cover and other 5 pages of *"Shanghai Hudec Architecture"* compiled by Hua Xiahong and Michelle Qiao on her Wechat Moments with an impressive comment, "No wonder, I like Hudec so much". I asked her, "Have you ever been to Hudec's Memorial Hall?" She answered, "Not yet, I don't know where it is." I was more than happy to accompany her to visit Hudec's Memorial Hall. Apart from this, I even asked her to give advice on how to hold Commemoration Month Events. Out of my expectations, she replied, "I need to read more books to learn more about Hudec." –To seek truth all the time is the quality of her, once the Chief Correspondent of Wenhui Daily.

At the same time, I thought, many effective works in various aspects have been done to introduce Hudec to the general public, but its coverage needs broadening and its depth needs furthering. –Is this not exactly the intention of holding "Love Home, Love Hudec" in "Shanghai Hudec Architectural Heritage Cultural Month"?

杜海军
Du Haijun

1978年出生于江苏宜兴，2003年毕业于中国美术学院，现为中国美术家协会会员。

曾获"全国第三届青年美展"优秀奖，"第十一届全国美展"优秀奖，"法国国际沙龙展"获金奖，多幅作品被中国美术馆、中国国家画院、中国油画学会、今日美术馆、辽宁美术馆等机构收藏。

Du Haijun, current member of China Artists Association, was born in Yixing, Jiangsu Province in 1978 and graduated from China Academy of Art in 2003.

Awards: Excellent Works in "the Third National Youth Art Exhibition", Excellent Works in "the Eleventh National Art Exhibition" and Gold Medal in "International Salon Art Exhibition, France".

Many of his works have been collected in China Art Museum, China National Art Gallery, China Oil Painting Society, Today Art Museum, Liaoning Art Museum and other institutes.

我的作品主要以城市楼群为主体，画中的城市似曾相识，但更多的是展现我心中的城市。记得十年前第一次来到外滩，被风格迥异的老建筑震撼了。岁月在墙体上留下了斑驳的痕迹，密密匝匝排列有序的窗户就像音符一样布满在高低错落的建筑上。这种形式感和色调正是我所想要的画面，于是就开始了以城市建筑为题材的创作。

我有很多作品都是以上海老建筑为创作原型，它们让我产生了创作冲动。尤其是建筑大师乌达克设计的建筑我最为喜欢。当年这位伟大的建筑师用一个外国人的眼光审视着这个多种文化交集在一起的东方大都市，他用他的建筑解读这个城市。今天我又在他的建筑中画入现代的人，用油画语言来重新解读。把建筑解构重组，采用一些现代化切割手法，画上纵横交错的线使画面形式感更强、更抽象。窗内的场景运用中国画的写意手法，把生活中的各种场景收纳在方形的小格子里。在创作时更多的是借用建筑的外部结构，把现代都市人丰富多彩的生活装进去。一个个小窗户就像安装在墙上的抽屉，收纳着每个家庭，装满了他们的生活；当夜幕降临时，窗内亮起一盏盏灯，就像垂直的建筑面上挂满了无数电视机，里面播放的是无休止的生活剧。

My works mainly focus on tower clusters in the city, among which some may look familiar to the viewers, but more resemble the city in my own heart. I still remember that I was shocked by the old buildings with a great variety of architectural styles when I first came to the Bund ten years ago. Time left the walls with spots and cracks. Windows crowded in rows like music notes written on buildings, low or high. The very form and hue in this scene is exactly what I dream of. So, I started painting urban architecture.

The archetype of many of my works is the old buildings in Shanghai. In other words, the old buildings has ignited my passion for painting, especially those designed by L. E. Hudec, which are my favorite. At that time, L. E. Hudec, the great architect, viewed Shanghai, an oriental metropolis, where various cultures converged with eyes of a foreigner. He demonstrated its cultural variety in his architecture. Today, I represent his architecture in modern paintings and decipher it via oil painting. I dissect and reconstruct the buildings with modern dissection techniques, and paint with crossed lines which have a clearer sense of form and look more abstract. I condense various scenes of life inside the windows into rectangular boxes in the traditional Chinese impressionistic style. I often instill the colorful life of modern metropolitans into the buildings. Each window is like a drawer in the wall which shelters a family and stores their life. When night falls, lights will be switched on inside the windows one by one so that the vertical façade resembles a scene with countless TV sets which play the unending life.

《城市轮廓 Contour of the City》 160×200cm 布面油画 oil on canvas 2013

《雨果故居 Hugo's Residence》 90×60cm 布面油画 oil on canvas 2015

《诺曼底公寓 Normandie Apartments》 120×80cm 布面油画 oil on canvas 2016

《四行储蓄会大楼 Union Building of the Joint Savings Society》 100×80cm 布面油画 oil on canvas 2016

《上海总商会旧址 Shanghai Chamber of Commerce》 73×54cm 布面油画 oil on canvas 2015

《后窗 Rear Window》 100×120cm 布面油画 oil on canvas 2012

《九号院的阳光 Sunshine on the Ninth Courtyard》 180×135cm 布面油画 oil on canvas 2011

李乾煜
Li Qianyu

1975年出生于上海，1999毕业于上海大学美术学院。现为上海视觉艺术学院美术学院教师、中国雕塑学会会员、上海美术家协会会员。

作品曾入选"第十二届全国美展""国际雕塑年鉴展""城市之光——国际雕塑大展""首届高校教师作品展"，并被鄂尔多斯雕塑公园、刘海粟美术馆等机构收藏。

Li Qianyu was born in Shanghai in 1975 and graduated from College of Fine Arts, Shanghai University. He is now a teacher of College of Fine Arts, Shanghai Institute of Visual Arts, a member of China Sculpture Institute and a member of Shanghai Artists Association.

His works have been presented in "the Twelfth National Art Exhibition", "International Sculpture Annual Exhibition", "Light of City — International Sculpture Grand Exhibition", and "the First Exhibition of College Teachers' Works". Besides, his works have been collected in Ordos Sculpture Park, Liu Haisu Art Museum and other institutes as well.

我对邬达克作品的理解是既现实又梦幻、变化多端、极富生命力且带有强烈的文化烙印。上海这座城市的独特魅力和邬达克作品的魔性是息息相关、密不可分的，我感受到他作品的深邃、大气、精致、诗意，但又有一丝琢磨不透、用语言不能言尽的气场。虽然他的建筑风格属于现代主义，但在东方文化语境里出现这种建筑语言着实令我们不得不再次审视艺术作品和艺术作品欣赏者的关系，以及在这两者关系背后早就存在的艺术真理。我试图用我的作品提出命题，与观众在某种连接下，透过东方的镜片共同寻求答案，向他老人家致敬！

In my mind, Hudec's architecture is both realistic and fantastic, very diversified in style, distinguished by great vitality and an explicit cultural mark. The unique charm of Shanghai and the unbreakable spell of Hudec's architecture are closely related and enhance each other. His architecture is profound, grand, exquisite and poetic, and meanwhile a little bit inconceivable and unspeakable. Although his architecture is modernistic, yet as this modernistic architecture is embedded in our oriental culture, we still shall reexamine the relation between architecture and its observer and the implicit truth of art behind this relation. I've tried to raise the previous proposition on his architecture, to search for an answer together with the observers from the oriental cultural background with my works to pay tribute to L. E. Hudec, the great immortal architect.

《时间抖了一下 Time Goes with a Swing》 190×50×30cm 核桃木 wood 2016

《时间抖了一下 Time Goes with a Swing》局部

《与邬达克共舞 Dance with Hudec》 250×320×40cm 橡木、铜板、霓虹灯 wood, copper plate, neon lights 2016

《与邬达克共舞 Dance with Hudec》局部

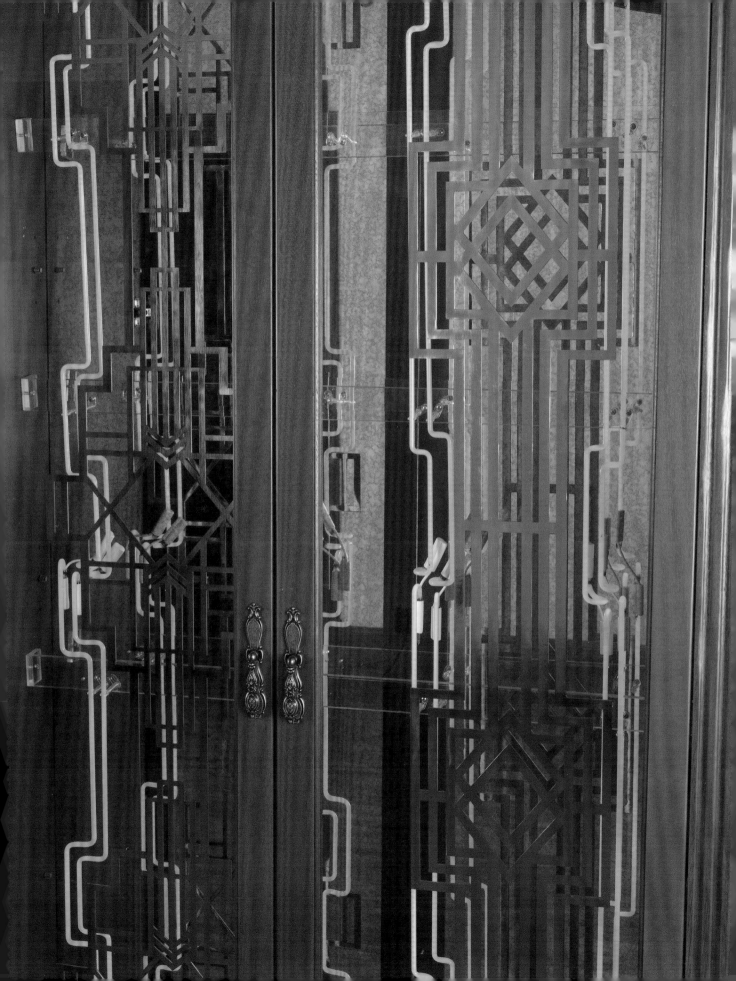

有故人的城池才是故乡

HOMETOWN IS THE PLACE
THAT HOUSED OUR FAMILY AND OLD FRIENDS

有故人的城池才是故乡

朱光

2016年4月19日，有一位来自瑞士的金发女士和她的先生一起来上海参观邬达克旧居，她说她和邬达克是叔侄辈。邬达克，匈牙利籍的斯洛伐克人，曾经作为奥匈帝国军队的一员参与战争，于1916年被俄国军队俘获，1918年辗转来到上海。来自建筑世家的DNA，让他为上海留下了近百幢建筑。1947年他离沪先是去了瑞士小住，后又定居美国。

这位来自瑞士的玛德琳女士，拿出了邬达克年青时的相片和邬达克写给她父亲的信件。原来，她是邬达克那一脉的后裔，她父亲在世时曾经跟她说起，家族里有一位亲戚在远东上海做建筑师，发展得很好，曾力邀她父亲也来上海。因为当时她父亲和母亲已经相恋，就放弃了来沪。近年上海持续不断的邬达克研究与宣传，在欧洲也引起了不少媒体的关注。2011年瑞士的报纸大幅报道邬达克与邬达克在上海的建筑设计。玛德琳女士不敢相信报纸上的邬达克就是她家族里的那位长辈亲戚。对照了家里收藏的邬达克照片之后，她才敢确认这位在上海取得巨大成功的邬达克正是她家族里的那位长辈亲戚。于是，玛德琳女士先找到了报纸上介绍邬达克及其作品的文章作者，再辗转来到上海，找到邬达克旧居修缮投资人刘素华女士，当时旧居刚开始修缮。玛德琳女士原本以为这个世界上只有她和她的兄妹姓"邬达克"，抵达上海邬达克旧居后，让她与失散多年的其他邬达克家族成员取得了联系。我见她的这次，已经是她第二次来上海了，也许这是她与上海冥冥中注定的缘分，她替她的父亲完成了上世纪邬达克发出的上海邀约。

有故人的城池才是故乡。玛德琳女士第一次来上海时，看到邬达克旧居修缮工地时激动不已，表示完工后一定要通知她。于是，这次她又来了，参观修缮完成的邬达克旧居，拿起了相机，把每一个角落拍摄给远在瑞士的亲眷们看。她并没有系统学习过建筑，却也饶有兴味地研究房屋的"建筑审美"，并在小本子上记下了许多邬达克在上海的传奇故事……

在她看来，那是她与已故父亲的隔空触摸，也是对已逝亲人伟大成就的亲眼见证。她从未想到，因为邬达克留在上海的建筑，使得自己也有缘上海，似乎上海也成为她半个故乡。

城池从来不是独立存在的石块与沙泥，她因为人们在改造石头与沙泥的过程中，被注入了情感，而成为人的创作。而越来越多的人的共同创作，使得我们扎根于此，安家落户。哪怕暂时远离，也会牵挂付出——付出的东西一旦传世了，此地就成为故乡。

表面看来，邬达克是个没有故乡的漂泊者。他出生于1893年奥匈帝国的兹沃伦州首府拜斯泰采巴尼亚。在第一次世界大战爆发后，他加入奥匈帝国的军队，还没挨过多少时间的军旅生涯，就被俄国军队俘房后送至西伯利亚战俘集中营。到了1918年11月11日，奥匈帝国年轻的卡尔皇帝宣布退位，结束了绵延7个世纪的哈布斯堡王朝的君主统治。邬达克身处"国破山河在"的境地。"祖国"奥匈帝国不久后分裂成奥地利、匈牙利、捷克斯洛伐克等。而邬达克流落到上海时，凭的是一份伪造的护照。后来，他才加入匈牙利国籍。此后，他才有了精神上的归属。身处上海，虽然他并不是犹太人，但是，他帮助了不少匈牙利籍的犹太人，这其中多少也有对"故乡"惦念的外在表现。他的职业黄金期，正与上海的城市崛起重合。在近三十年里，他吸取当时国际建筑风潮里的 Art Deco、Bauhaus 等，也复兴了哥特式、都铎式等欧洲古典审美意趣，为上海的"万国建筑博览"添上了浓墨重彩的一笔。

最值得记取的不仅有他多样化的审美情趣为上海这座城池带来的丰富视觉效果，还有他在设计国际饭店时的科技创新，让人们最终敢于在上海这个冲积型平原上建造摩天大楼。上海是泥沙冲积而成的，横沙岛、崇明岛，至今还会因为泥沙冲积而"成长"。在这样的泥沙地上如何建造摩天大楼？起初他想建国际饭店的计划，并未得到欧美同行的认同，他们认为邬达克是在异想天开。结果，他创造性地使用了筏式地基，开创了建筑史上软土地基上建造摩天大楼的先河，同时国际饭店的建成，也成就了邬达克软土地基上建造摩天大楼的鼻祖地位。在以后的半个世纪里，80多米高的国际饭店始终是亚洲第一高楼。

如今，在徐汇、黄浦、长宁一带，转角就能遇到手持单反相机的男女文青，他们未必知道筏式地基，但是一定会沉迷于在国际饭店、大光明电影院、沐恩堂、吴同文住宅、中西女中、武康大楼等处里里外外留影。他留在上海的诸多建筑，构成值得我们怀恋的老上海的标识。建筑，能令人想起另一个时空里的唯美。如果说，现实生活中能有什么可以形成令人穿越的时光机，那就是建筑——凝结着另一个时代的技术与艺术。可惜，上海只是邬达克作品的故乡，他一生的建筑作品几乎都留在了这里。1947年赴美定居之前，他曾经离沪赴瑞士小住，这就是他与玛德琳女士生命轨迹的交集，哪怕当时这位来沪寻访的瑞士玛德琳女士还未出生。定居美国加利福尼亚之后，因山体滑坡邬达克失去了建在海边山崖的花园别墅，接着邬达克为自己设计了瑞士风格的木屋，可惜未入住，就于1958年因心脏病去世，始终没有了却回归故乡的心愿。

在这世事变迁、时局动荡、茫茫人海、前途未卜的尘世里，邬达克的故乡在哪里呀？他把他的精神财富都留在了上海。从这个意义上而言，上海，应该也可以算做他精神上的故乡。时至今日，举办纪念邬达克绘画雕塑邀请展，其实，也是对邬达克精神故乡以及上海多样化的城市线条的描摹。

建筑与绘画是姻亲。绘画能使得建筑流动起来、传播各处，却又有其本身的风韵。人的创造，使得建筑与绘画都蕴含了情感、遐思和永久的生命力。因而，哪怕斯人已去，人的创造物，不仅仅是建筑、绘画，只要是作品，都可以成为他们的精神故乡，值得后人细细品味，从中感受到故人对城池的深情，给予了我们绵长的滋养，足以传世。

Hometown is the place that housed our family and old friends

By Zhu Guang

On April 19th, 2016, a Swiss blonde came to Shanghai with her husband to visit Hudec's Residence. She told me that Laszlo Hudec is her uncle. Laszlo Hudec is Slovakian Hungarian. He served in the military of Austro-Hungarian Empire in WWI, and was captured by the Russian army in 1916, and finally found his way to Shanghai. He, born into a family of architects, left nearly one hundred buildings in Shanghai. In 1947, he left Shanghai for Switzerland where he stayed for a while, and then settled down in the USA.

This lady is Mrs. Madeleine. She showed me photos of young Laszlo Hudec and the letters Laszlo Hudec wrote to her father. I was aware of the fact that she is a descendent of the Hudec family. Her father once told her that a family member of his worked as an architect in Shanghai in the Far East with promising business there, and enthusiastically invited him to Shanghai. But at that time, her father was in love with her mother, so they decided against it. In recent years, many researches and publications on Laszlo Hudec, her uncle keep coming out in Shanghai, which attracts the attention of many news agencies in Europe. In 2011, a Swiss newspaper reported at length Laszlo Hudec and his architecture in Shanghai. Mrs. Madeleine could not believe that the Laszlo Hudec on the newspaper is actually an uncle in her family. After comparing the photo on the newspaper with the photos her father had left her, she was ascertained that they were the same Laszlo Hudec, who had made it in Shanghai. So, she contacted the author of that article, and then came to Shanghai with the information provided by the author and found Lady Liu Suhua, the investor for the renovation of Hudec's Residence when the renovation just got under way.. Originally, Mrs. Madeleine thought that all over the world only she and her brothers and sisters inherit the family name "Hudec". After visiting Hudec's Residence, she got the opportunity to get in touch with other members of the Hudec family. It was her second time to Shanghai when I met her. Perhaps this meeting in Shanghai is predestined by the Almighty. She came to Shanghai at the invitation sent by her uncle Laszlo Hudec to her father last century.

Hometown is the place that housed our family and old friends. When Mrs. Madeleine came to Shanghai for the first time, she was very excited to see the renovation field of Hudec's Residence, and requested to be notified of its completion. So, she came again, and raised a camera to take photos of each corner of the renovated Hudec's Residence which she will show to her family members in Switzerland. Although she has no systematic knowledge of architecture, she was greatly interested in " architectural aesthetics" , and took many notes of the legend of Laszlo Hudec on a small notebook…

In her opinion, that is a special contact with her late father, and a witness of the great achievements of her family member who had passed away. Quite unexpectedly, the buildings Laszlo Hudec had left in Shanghai allowed her an opportunity to come to Shanghai. Shanghai seems to be a hometown of sort for her.

A town is never merely a pile of stones, sand and mud. A town is the creation of the people for they have instilled their love into it in the process of working with stones, sand and mud. More and more people join in the creation so that we could settle down in the town and live here with our family. Even if we are away from the town for just a while, we still care about the work we have done for the town—when the work passes on for generations, the town then will become our hometown.

 Laszlo Hudec is seemingly a homeless drifter. He was born in Besztercebánya, the capital of Zvolen State, Austro-Hungarian Empire in 1893. When WWI broke out, he joined the military of Austro-Hungarian Empire. But after short service in the military, he was captured by the Russian Army and was sent to Siberian POW Concentration Camp. On Nov. 11th, 1918, Karl I, the young emperor of Austro-Hungarian Empire, declared his resignation, showing the end of the monarchy of the seven-century-old Habsburg Dynasty. His homeland was then in "disintegration". No sooner, his homeland, Austro-Hungarian Empire disintegrated into Austria, Hungaria and Czech-Slovakia. When Laszlo Hudec drifted to Shanghai, he registered his identity with a fake passport. Later, he obtained his nationality as a Hungarian that provided him with a spiritual homeland. When in Shanghai, he helped many Jewish Hungarians although he is not a Jew, which somewhat expresses his "concern" of his homeland. The heyday of his career was in tune with the rise of Shanghai. For nearly thirty years, he had absorbed Art Deco and Bauhaus in the international trend of architecture at that time and revitalized the classic aesthetic flavor of Gothic and Tudor architecture in Europe, which contributed an essential scene for "Exotic Building Clusters" in Shanghai.

What's most memorable is not only a variety of aesthetic flavors that he brought in for Shanghai with splendid visual effects, but also the technical innovation that he applied to the design of Park Hotel. Thanks to his trend-setting innovation, people finally were encouraged to build a skyscraper in Shanghai, a city on the frail alluvial plain. Shanghai is a region formed with sediments from floods. Hengsha Island and Chongming Island are still "growing" because of coming sediments now. How to build a skyscraper on the sediment soil? Initially, his proposal of Park Hotel was disapproved of by his compeers from Europe and America. They thought, Hudec was just fancying. Consequently, he creatively applied raft base for it, which initiated the construction of skyscrapers with a raft base on soft soil in the history of architecture. It is the completion of Park Hotel that won him "the Father of Skyscrapers on Soft Soil Base". In the coming half a century, the 80-meter-high Park Hotel remained the highest building in Asia.

Now, when you walk in Xuhui, Changning or Huangpu, in each corner, you will find young ladies or gentlemen with a single-lens reflex camera. They may not necessarily know the raft base, but must be fascinated by Park Hotel, Grand Theatre, Mu'en Church (former Moore Memorial Church), Residence of D. V. Woo, McTyeire School, Wukang Building (former Normandie Apartments), and will take pictures inside and outside of these buildings. Many buildings of his are now the marks of Old Shanghai that we still miss very much. Architecture reminds us of the beauty existent in the past. If there is Time Machine that can take man to travel back through time in the real world, then the Time Machine shall be architecture which materializes technology and art in another age. The pity is that Shanghai is only the hometown of Hudec's architecture. Nearly all of his works are left here in Shanghai. In 1947, before he settled down in the USA, he left Shanghai and stayed several days in Switzerland when he had involved with the life of Mrs. Madeleine, even if she was not born yet then. When he settled down in California, the USA, he lost one garden villa on the cliff ridge because of landslide. Then he designed a Swiss wooden dwelling. Unfortunately, he died of heart disease in 1958 before he moved in. In the final analysis, he still failed to return to his hometown.

Time is gone. The world has changed. The future is unpredictable. It is hard to find one person who lived in the past. Then, where is Hudec's hometown? He has left all his spiritual wealth in Shanghai. In this sense , Shanghai shall be taken as his spiritual hometown. Today, a painting and sculpture invitational exhibition is held here in memory of him—Laszlo Hudec. Actually, this exhibition is also a depiction of Hudec's spiritual hometown and the diversity of Shanghai.

Architecture and painting are in-laws. Painting gives legs to architecture so it can go around the world. And also painting has the charm of its own. It is people's creation that fills love, fancy and immortality into architecture and painting. Therefore, even if the creators pass away, their creation, be it architecture or painting, will be the hometown of their spirit that is cherishable to the future generations. We can feel their deep love for the town. And their love nurtures our lives. That is why their creation will be passed on to the future generations.

洪 健
Hong Jian

1967年出生于上海，1991年毕业于上海大学美术学院，2002年结业于上海中国画院首届高级研修班。现为上海中国画院画师、展览部副主任。中国美术家协会会员、上海市美术家协会理事。

曾获"第五届上海美术大展""白玉兰美术奖"一等奖，"第十一届全国美展"银奖，上海市文联"优秀中青年艺术家"称号，作品被中国美术馆、上海美术馆、上海中国画院等机构收藏。

Hong Jian was born in Shanghai in 1967. He graduated from College of Fine Arts, Shanghai University in 1991 and completed his courses in First Advanced Training Class of Shanghai Chinese Painting Academy in 2002. He is now a painter and the deputy director of Exhibition Department in Shanghai Chinese Painting Academy, a member of China Artists Association and a director of Shanghai Artists Association.

Awards: No.1 Magnolia Award for Fine Arts in "the Fifth Shanghai Grand Art Exhibition", Gold Medal in "the Eleventh National Art Exhibition", the title of "Outstanding Young and Middle-aged Artist" by Shanghai Federation of Literary and Art Circles. His works have been collected in China Art Museum, Shanghai Art Museum, Shanghai Chinese Painting Academy and other institutes as well.

记得美院附中时期，常年在市区画风景写生或是干脆骑着自行车自得其乐地闲逛，国际饭店、大光明电影院、沐恩堂、美国总会、爱司公寓等建筑自然熟悉得很；也记得在美院附近见识了达华公寓、市三女中；还记得迁入徐汇后，饭后散步在诺曼底公寓周边或是交大工程馆，蹒跚学步的小女也曾跌跤在诺曼底公寓的廊下狂哭不已；更记得几个曾经工作的处所附近还有爱神花园、巨鹿路花园住宅、白公馆等……

　　无处不在的邬达克建筑始终围绕在身边。我自始至终觉得传奇的上海造就了传奇的邬达克，传奇的邬达克与传奇的上海相遇，这才是真正的上海传奇故事。

I still remember, as a student in the High School Affiliated to Shanghai Academy of Fine Arts, I often went out painting the scenery, or merely enjoyed myself bicycling across the streets. Certainly, I was very familiar with Park Hotel, Grand Theatre, Moore Memorial Church, American Club, Estrella Apartments and other buildings. I remember on my bicycling tour around the city I saw Hubertus Court and No.3 Middle School for Girls (formerly Mc Tyeire's School for Girls) near Shanghai Academy of Fine Arts and thus my visions were broadened.

 I remember that later my family moved to Xuhui District where after dinner we would have a walk around Normandie Apartments or Engineering and Laboratory Building of ChiaoTung University. Once My little daughter toddled and fell down in the corridor of Normandie Apartments and kept wailing for a long time there.

Of course, I remember all the more clearly that near my former offices there are Garden of Love, Residential Garden on Julu Road, and Residence of General Bai Chongxi, etc.

Hudec's buildings are everywhere around us. I always think that it is legendary Shanghai that made Hudec a legendary success and when legendary Hudec met legendary Shanghai, a true Shanghai legend was created.

《绿房子 Green House》 43×45cm 纸本设色 color on paper 2016

《孙科别墅 Sun Ke's House》 67×67cm 纸本设色 color on paper 2016

《上海故事 No.1 Stories of Shanghai No.1》 48×46cm 纸本设色 color on paper 2013

《邬达克哥伦比亚生活圈一隅 A Corner of Columbia Community》 51×51cm 纸本设色 color on paper 2013

《上海故事 No.2 Stories of Shanghai No.2》 63×63cm 纸本设色 color on paper 2011

《上海故事 No.3 Stories of Shanghai No.3》 33×33cm 纸本设色 color on paper 2011

《雷雨季节——马勒别墅 Stormy Season—Moller Villa》 96×96cm 纸本设色 color on paper 2010

《从苏河远眺真光大楼 From Suzhou River overlooking Zhenguang Building》 68×68cm 纸本设色 color on paper 2012

《春水向东·苏河 No.1 A Spring Stream Flows to the East · Suzhou River NO.1》 68×68cm 纸本设色 color on paper 2013

《春水向东·苏河 No.2 A Spring Stream Flows to the East · Suzhou River NO.2》 51×51cm 纸本设色 color on paper 2014

毛冬华
Mao Donghua

1971年出生于上海，1995年毕业于上海大学美术学院。曾任刘海粟美术馆研究部副主任兼展览部负责人，现为上海大学美术学院附中校长。中国美术家协会会员、上海美术家协会会员、上海青年文联美术专业委员会副干事长。

曾获"第六届全国体育美展"优秀奖，"上海青年美术大展"三等奖，"第七届上海市美术大展"白玉兰美术奖优秀奖，"第十二届全国美展"提名奖；获上海市"园丁奖"、上海市"文化新人"等称号。作品被中国美术馆、中国体育博物馆收藏。

Mao Donghua was born in Shanghai in 1971 and graduated from College of Fine Arts, Shanghai University in 1995. She was once the deputy director of Research Department and the director of Exhibition Department in Liu Haisu Art Museum. She is now the principal of the High School Affiliated to College of Fine Arts, Shanghai University, a member of China Artists Association, the deputy secretary general of Fine Art Expert Committee of Shanghai Federation of Young Writers and Artists.

Awards: Excellent Works in "the Sixth National Sports Art Exhibition", Third Prize in "Shanghai Art Exhibition for Young Artists", Magnolia Award for Excellent Art Works in "the Seventh Shanghai Art Exhibition", Award Nomination in "the Twelfth National Art Exhibition", Shanghai Gardener Award, Shanghai New Talent in Culture and other titles as well. Her works have been collected in the National Art Museum of China and China Sports Museum.

我在建筑题材的绘画上，已经进行了一段时间的尝试。刚开始是一种与新老建筑之间对话的过程，同时对柔性的水墨表现建筑的质地比较敏感。现在则更注重绘画本体语言的探索，回归到笔、墨、纸这个原点。具体实践以主要运用花卉用笔表现建筑，改为主要运用山水画的技法；以注重呈现建筑本体改为注重呈现笔墨、建筑作为载体研究探索绘画本体语言的丰富性及厚重感。其次，更注重创作前的现场体验，感受建筑在立体空间里的体量和气场以及每栋建筑的姿态和性格。根据天气和光线的不同，同一题材创作多幅作品。运用中国绘画的留白产生的光感来营造情景和气氛，也是我创作实践的一个课题。第三，提高对媒材特点的掌握，如以前选用陕西镇巴宣纸，它适合一次完成没骨花卉笔法，而今多用安徽宣纸，它适应积墨复笔的山水没骨技法、力求水、墨和宣纸的完美结合。

　　在绘制《武康大楼》的时候，我在实地考察过程中发现武康大楼沿淮海中路的这一面虽为人熟知，但构图会显得比较平，而它的背侧面有两块凹立面，形成了"隔断"，与其他平面构成一种节奏感，于是选择了背侧面这个角度。这幅画，我以没骨法画成，线条既不能显得粗糙，又要维持一种灵动感，每画完一段密集的砖墙窗格，凹面正好成了休止换气之处，如此再继续……留白和透气之处早在经营布局中就想好了，总的来说这幅画画得比较顺。画小白楼的时候，机缘巧合，我获得了它早年刚建成时的珍贵档案照片，这引起了我极大的兴趣想去重现它的原貌。建筑也是一个文脉，看到原貌人们更能理解过去，体会现在。

I have been experimenting with painting architecture for quite a long time. In the very beginning, I was very sensitive to the contrast between old and new buildings presented in soft ink. But now I focus more on the ontology of painting and return to the basic elements—Chinese brushes, ink and paper. In practice, I mainly apply the method of painting landscape instead of the method of painting flowers to present architecture. My focus is more on brushes and ink than on architecture itself. Architecture is just a carrier of painting elements which demonstrates its richness and intensity. Second, I spend more time on observation before painting a building. I would feel the volumn and the energy field of cubic space in architecture, its posture and character. I would paint several works for the same building under different whether and illumination. I create a special scene or atmosphere with the effect of vacancy in traditional Chinese painting, which has been a subject in my painting practice. Third, I've improved my skills with painting media. In the past, I usually painted on Shaanxi Zhenba paper which is suitable for a boneless painting of flowers finished once and for all. But now I paint on Anhui Xuancheng paper which is suitable for a boneless painting of landscape added to and polished and re-polished later. I try my best to create the perfect combination of water, ink and paper.

When I was painting "Wukang Building"(once known as "Normandie Aparments"), I found in my field trip that although the façade of Wukang Building facing Middle Huaihai Road is familiar to the people, its view will look too plain, while its back has two concave façades which form a "partition" and create a change of rhythm distinct from other façades. So, I chose to draw its back. I painted this picture with boneless painting method, in which the lines were kept not too bold, but flexible enough. After painting a close row of windows in the brick wall, the concave served as a very transition. Then I would go on drawing another row of windows…… Vacancy and ventilation were already reserved in my mind. Generally, the whole process went on smoothly and successfully. When I was painting the little white building beside Wukang Building, coincidently, I got the precious documentary photos which were taken in the early years after Wukang Building was just unveiled. These photos aroused my great interest in recovering its original appearance. Architecture is also a cultural context. When the people see its original appearance, they will understand the past and the future deeper.

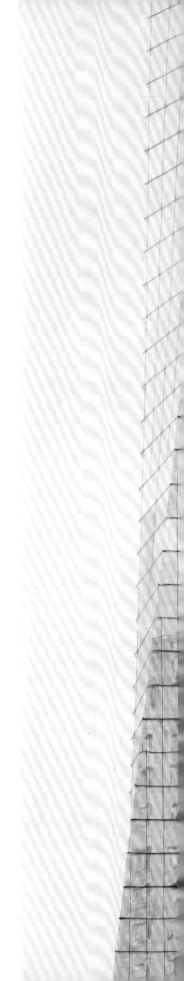

《窗里窗外 In and Out of the Window》 120×120cm 纸本水墨 ink on paper 2011

《小白楼——上海工艺美术研究所 White Villa》 70×85cm 纸本水墨 ink on paper 2015

《邬达克旧居 Hudec's Residence》 68×68cm 纸本水墨 ink on paper 2016

《可以远眺的房子 The Height for Far Sight》 68×68cm 纸本水墨 ink on paper 2015

《武康大楼 Wukang Building》 68×68cm 纸本水墨 ink on paper 2014

《外滩二号 No.2 Bund》 68×70cm 纸本水墨 ink on paper 2014

《外滩七号 No.7 Bund》 68×68cm 纸本水墨 ink on paper 2016

《记忆系列——字典 Memory – Dictionary》 52×52cm 纸本水墨 ink on paper 2015

《记忆系列——打字机 Memory – Typewriter》 68×45cm 纸本水墨 ink on paper 2016

《冬日暖阳 No.1 Sun in the Winter No.1》 98×90cm 纸本水墨 ink on paper 2010

《冬日暖阳 No.2 Sun in the Winter No.2》 98×90cm 纸本水墨 ink on paper 2012

应海海

Ying Haihai

1955年出生于上海，1982年毕业于上海师范大学美术学院，现为上海师范大学美术学院副教授、硕士研究生导师。中国美术家协会会员、上海美术家协会会员。

曾获"全国小幅油画展"优秀作品奖，"上海水彩粉画展"优秀作品奖，作品被藏家收藏。

Ying Haihai was born in Shanghai in 1955 and graduated from College of Fine Arts, Shanghai Normal University in 1982. He is now an associate professor, MA supervisor in College of Fine Arts, Shanghai Normal University, a member of China Artists Association and a member of Shanghai Artists Association.

Awards: Excellent Works in "National Small-size Oil Painting Exhibition" and Excellent Works in "Shanghai Watercolor Painting Exhibition". His works have been collected by many collectors.

我们所生活的这座城市，是一个充满无穷魅力的地方。它的过去是那么出色地融合了中西文化，创造出远东最为繁华的城市景观，见证了整个中国近代的发展和变化，为我们留下了许多经典的老建筑和街景，也留下了许多让人回味无穷的文化历史。每每徘徊在黄昏下，那些老建筑总让我感动又让我思考，当栉风沐雨的斑驳墙面和屋顶披上了金色的阳光时，飘忽跳跃的光影中依稀可辨的是恍惚而过的历史尘影，是逝去了的旧日梦境，还是现实？我们在这些饱含历史的老建筑里寻觅过去的风云变幻，也在憧憬中回恋这份缠绵悱恻的记忆。

其实这些老房子的价值早已超越了建筑本身，它正无声地叙述着世事沉浮和沧桑巨变。当我们关注这座城市的今天，势必会更加怀念那些已经消失或者正在消失的风景。这些老建筑里，既有一座城市所留下的历史脉络，同时也有父辈们和我们所留下的生活印迹和奋斗经历，包含着我们的欢乐和苦恼、希望和梦想，使人永远无法忘怀。

我就是带着这样一份心情来描绘邬达克的建筑。

Shanghai, where we live now, is a city full of endless glamour. Shanghai converged Chinese and western culture outstandingly, created the most flourishing civic landscape in the Far East, witnessed the development and progress of modern Chinese history, and provided us with many classic heritage buildings and streets, and many durable cultural stories appealing to us. Whenever I wander in the streets at sunset, the old buildings never fail to touch me and make me think. Whenever the aged walls and roofs are aglow in golden sunshine, I cannot help myself wondering what we really see in the fleeting light. Is it a shadow of our past just passing by? Is it the old dream that comes true again? Or is it the very reality? I search for the turbulent past from the buildings full of stories and recall the sentimental memories of love.

Actually, the real value of these old buildings transcends the buildings themselves. It lies in their silent narration of the ups and downs, twists and turns in our past. If we really care about present Shanghai, we will inevitably yearn for the scenes that have disappeared or that are disappearing. The old buildings carry the historic traces of a city and the footprints of our striving fathers, contain our laughter, distress, hope and dream, which are all engraved on our memory.

That is why I paint Hudec's buildings.

《沐恩的阳光 Sunshine on Moore Memorial Church》 50×60cm 布上油画 oil on canvas 2016

《邬达克旧居 Hudec's Residence》 80×100cm 布上油画 oil on canvas 2016

《温暖的午后 A Warm Afternoon》 60×80cm 布上油画 oil on canvas 2015

《晨曦中的绿屋 Green House in the Morning》 50×60cm 布上油画 oil on canvas 2016

《江苏路 Jiangsu Road》 80×100cm 布上油画 oil on canvas 2010

《上海南市 Nanshi District, Shanghai》 80×100cm 布上油画 oil on canvas 2012

《上海弄堂 No.1 A Lane in ShanghaiNo.1》 140×110cm 布上油画 oil on canvas 2009

《上海弄堂 No.2 A Lane in Shanghai NO.2》 80×60cm 布上油画 oil on canvas 2015

《上海弄堂 No.3 A Lane in Shanghai NO.3》 130×150cm 布上油画 oil on canvas 2014

《上海印象 The Impression of Shanghai》 80×100cm 布上油画 oil on canvas 2013

为一座城市造房子的人
THE MAN WHO BUILT HOMES FOR A CITY

为一座城市造房子的人

毛时安

　　他是一个幽灵。在我生活的这座现代化大都市里，如水银泻地无处不在。他像长笛透明的音符潇洒地洒落在湖畔河边弄堂深深的尽头，像气势磅礴的交响曲轰响闹市宽阔的马路。你可以不知道他的名字，不认识他的人，但你无法避开他，因为不经意之间你就会在这座城市里遇见他。他把他的灵魂、他的才华镌刻在了这座城市的土地上。然后他就一去不归了，消失在一片茫茫的虚无之中。

　　六十年前，我还是个懵懵懂懂少不更事的孩子。那时弟弟妹妹还没有都出生，家境也还过得去。一年过生日，父母带我走进南京西路头上一栋深棕色的高楼。它的雄姿矗立在路边，是那个时代上海的地标，是最高的建筑。在周围楼群的簇拥下，它像帝王那么的尊贵显赫，那么的一览众山小。就像姚明站在我们中间。大家习惯的用上海话亲切而尊敬地叫它"廿四层楼"。父母带我在三楼的窗边坐下。大街上铺满了金色、透明的阳光。南京路永远川流不息的人群和各式各样的车子，隔着同样透明的一直顶到天花板的巨大窗玻璃，在阳光里自由地移动。不时，还有拉着铁栅栏门的1路有轨电车，拖着长长的"辫子"，叮叮当当慢悠悠地驶过。父母给过生日的我买了一本《小朋友》，一个长着很长白胡子的消瘦老人在雪白的书页里微笑。我看懂了，一个叫齐白石的画家得了国际和平奖。他的伟大，我一直要到很多年以后才明白。但他画的那些虾真和活的一样，在看不见的水里游动，感动了一个孩子的一生。于是，我记住了一座叫国际饭店的建筑，一座近半个多世纪雄踞亚洲高度宝座的建筑。旁边还有一座浅黄色横在南京路上，隔着黄河路，似乎在和国际饭店说悄悄话的建筑。建筑在一片片连续的薄板中升起风帆般地隆起，写着"大光明影院"几个大字，入夜，霓虹灯在夜空骄傲地闪烁。二十世纪八九十年代，我一度和电影走得很近很近、很热络。在那里，围着白杨、张瑞芳、孙道临、秦怡，还有谢晋、吴贻弓，指点中国电影如画江山。还有张艺谋、陈凯歌、吴子牛许多若干年后如日东升的电影艺术家们坐而论道，目睹了中国电影的又一段一去不复返的红红火火的流金岁月。

　　后来，我像做梦一般跨进了童年向往的圣殿般的作家协会工作。开始有人介绍说是旧社会"火柴大王"刘鸿生的宅院。过了一阵子，有人纠正，是他弟弟刘吉生的府邸。绘着风景的长条彩色镶嵌玻璃窗，神秘美丽的光亮中，旋转扶梯从一楼盘旋到三楼，扶梯的铸铁栏杆上"KS"的纹饰，证实了主人的信息。从楼梯顶部往下，可以看见楼梯像海贝一般美丽的圆弧袅袅升腾而起。底层黑白相间的地坪，像数学方程式般简约明快。我在小楼西南角的204房间办公，一个房间里年长的有我老师徐中玉、诗人罗洛，还有同龄的赵长天、宗福先和叶辛。房间南面有铸铁的大门，门两边有连着墙体嵌着大镜子的欧式雕花水曲柳衣柜。推开大门，就可以看见院子里的花园，不管世事变迁，永远郁郁葱葱的绿。那尊由巴金老花匠李师傅"文革"中保下来的女神普绪赫大理石雕像伸展着美丽婀娜的半裸躯体，仰望着蓝天。几个娃娃和一汪清水，围绕着她。我们站在阳台上，有时独自一人，有时和长天一起，伸手几乎可以摸到撑起小楼的希腊爱奥尼式的圆柱。爬山虎的藤蔓和枝叶婆娑地沿着深棕色的墙头不依不饶地伸过来。小楼的东边是《收获》编辑部。巴金先生的女公子李小林和她的编辑们永远地埋在来稿堆里。编辑部房间外面有一个阳台，它的下面同时也是进作协小楼的必经之地。作协成立40周年，为了拍电视片我和赵长天、陆星儿、彭瑞高、竹林一起拉开嗓子高唱过《年轻的朋友来相会》，其实那时我们都已四十出头了。204房间是当年作协所有事情的开端，无论对的还是错的，现在一切都远去了，连同当年主人家几个房间里发生过的各种惊险浪漫传奇的故事。唯有房子，总屹立在院子里。

在文化局工作，逢年过节我去看望老艺术家孙道临、王文娟夫妇。淮海路和武康路夹角，一栋七层的高楼拔地而起。从上面看，像一只巨大的熨斗。从正面看，像一艘起锚远航乘风破浪的巨轮。这座楼使周围一切猛地就有了庄重恢弘优雅的不凡气度。七楼的环形大阳台，可以极目远眺，似乎能把人间万象尽收眼底。整栋楼就像马勒的交响乐，宏大而不失次序。沿街长长骑楼下的连续券门，就是一小节、一小节由小号吹奏出来的嘹亮乐句。孙道临、王文娟夫妇就住在这栋大楼后来加的四楼里。当年赵丹、王人美、上官云珠、秦怡、郑君里这些中国家喻户晓万众瞩目的电影明星都曾在这里过过"日脚"。在孙道临、王文娟的面前，我真正领略了气质和风度。大楼曾有过一个显赫威风的名字：诺曼底公寓。现在很平民，武康大楼。还有，我写过评论的《蓝屋》的原型绿房子，染料大王吴同文的宅邸。在华东医院老楼，我闻到了一股混合着来苏水味道的肃穆气息，在走廊间飘浮……雁过留声。这座城市星星点点到处回响着他留下的声音。那是建筑，凝固的音乐。

恕我孤陋寡闻，在很多年以后我才知道，这些建筑出自同一位建筑师的手笔。他就是当今已经赫赫有名的邬达克——一个来自欧洲的建筑师，一个连国籍是匈牙利还是斯洛伐克都不太分明、身上却流淌着两国血脉的建筑师。就是这样一个建筑师，在上海，他居然留下了住宅、影院、教堂、医院、饭店大大小小近百件建筑作品。其中三分之一列入上海优秀历史建筑。它们不娇不宠却流光溢彩，装点着这座城市。就建筑师和上海这座城市的关系而言，迄今无人能出其右。以致建筑史家如此断言，如果没有邬达克，上海的建筑史将完全是另一部建筑史，而且上海的近代建筑史将不得不重写。这种情景，在人类建筑史上实属罕见。

就建筑师和城市的关系的重要性而言，邬达克使我想起西班牙巴塞罗那的伟大建筑天才高·迪。一座和上海同样伟大的城市。虽然20世纪最伟大的现代艺术家毕加索、米罗、达利都和巴塞罗那有着千丝万缕的联系。但走在巴塞罗那街头，我最大的感动居然不是他们，而是，那个叫"高·迪"的建筑师。他的旖旎瑰丽像来自云端的奇思妙想，飘忽在巴塞罗那街头的屋顶瓦檐烟囱砖墙门楣窗台上。米拉公寓、巴特拉公寓、古埃尔公园……它们像海浪一样的曲线汹涌澎湃地涌来，室内像大海里的贝壳盘旋蜿蜒，充满了自然主义神秘诡异而又生动的气息。圣家族大教堂170米高的尖顶犹如巴比塔直刺蓝天，又如一部结构宏伟迷宫般的交响乐，永远被脚手架和塔吊簇拥着，一百多年了，建设还在"正在进行时"中。在巴塞罗那，你似乎走不出高迪的怀抱，就像我一个穷人的孩子，在上海，还不懂事，就被邬达克拥抱过了。

让人感兴趣的还有两人虽有不同但却同样奇异的人生。邬达克的传奇在于，一个一战战俘营逃出来的战俘，穿过风雪弥漫的西伯利亚大平原，颠沛流离来到远东。在一个举目无亲的大都市留下了一批熠熠闪光的"房子"，成了这座城市青史留名的建筑师。他一定很富有，梦里不知身是客，直把他乡作故乡。在一个战乱不断的时代，连国籍身世都无法确定，没有了栖身的祖国，他的心恐怕是一直有点支离破碎，甚至千疮百孔的。这种隐隐的痛，真是无人诉说的。从1918年孑然一身踏上外滩，到1947年举家飘然而去。他像一颗划过夜空的流星，一瞬间留下了耀眼的灿烂。他离别得那么决绝，此后，除了自己栖身的居所，再没有留下一座建筑！最后，心肌梗塞，客死他乡，结束了65年的生命旅程。高·迪性格乖张古怪落落寡合，三年五年天天同一套衣服，又脏又破，终生未娶。最后，被通车典礼上披着彩旗的有轨电车撞倒。人们把这个糟老头视为乞丐，送到医院就咽了气。唯有一个老太，后来认出了他。高·迪终年74岁。全城的百姓涌向街头，为他送葬。他们把所有的才华毫不吝啬地挥洒给了一座城市。作为建筑家，他们当得起后人的尊敬和赞美。我们这些庸人各有各的庸俗，唯有天才的命运都是相似的。

《爱屋·及邬》里的这些画家，我大都很熟悉。他们生活穿行在这座城市的大街小巷，致力于在宣纸和画布上描绘这座城市的肖像，用原创的发自内心的艺术语言。这座城市已经成了他们心头永恒的爱人。她绰约的风姿，内心的律动，寒暑晨昏的表情，乃至她的忧伤，他们都了然在心。展现在他们的作品里，和现实一样的清晰，又像梦境一般的遥远飘渺。他们真的是爱这座城市，爱这些房子，爱设计这些房子的邬——达——克。

对于我来说，终于有一天，来到番禺路弄堂深处129号的邬达克旧居。这栋1930年落成的都铎式乡村别墅，在当年西郊大片田野的衬托下，明木的结构和雪白的墙体，色彩明朗脱俗，曾经优雅清新得像飘然而来的简爱，亭亭玉立。在经历了大半个世纪的风雨洗礼后，门窗脱落，墙面翘裂神情呆滞,像一个风雨中的弃儿。这座曾经风光的老房子,已经衰败凋零了。同是异乡人,一个在商海沉浮多年的湖南女子居然在茫茫人海中一见钟情地瞅上了她。就像一个痴心的女孩为了心爱的人,不惜飞蛾扑火,甘愿为此献出一切。像抢救一个奄奄一息的婴儿,她克服了各种难以想象的困难,1000多个日日夜夜的心血洒在了这座老房子的每一个角落。我曾经在她陪同下,从一楼到三楼,走进每一个房间。人去楼空,冥冥中,往昔的岁月像月光下的晚潮无声无息地汹涌而来。主人一家围坐在大厅餐桌边的欢笑在吊灯下回响。21世纪邬达克旧居,焕发了昔日曾经的姣好容颜。

  入夜,在扶疏的林木间灯光摇曳。很温暖。有飞鸟永远停在了尖尖的屋顶上。那是当年主人设计的飞鸟雕塑。难道那是主人灵魂的归来?停留?这个为上海造房子的人啊。

THE MAN WHO BUILT HOMES FOR A CITY

By Mao Shi'an

He is a phantom. In this modern metropolis where I live now, he is omnipresent, just like the mercury dispersed onto the ground. He is like music notes floating from the flute, falling freely on the edge of a lake, the bank of a river or into the far end of a lane. Or rather he is more like a piece of overwhelming symphony roaring on the broad busy streets. You may not have heard of him; you may not have ever come across his name for even once in your life, but you cannot avoid him, for you will encounter him in this city. He had engraved his spirit and his talent on the soil of this city. And then he was all the way gone and vanished into thin air.

Six decades ago, I was an innocent boy. My younger brothers and sisters were not born yet. My family was quite well-off. I remember once on my birthday, my parents took me to a deep brown skyscraper at the end of West Nanjing Road, standing there with a steeple and heroic posture. This skyscraper was the highest building and the landmark in Shanghai. Surrounded by clusters of dwarf buildings, it looked like a majestic and eminent king watching down his subjects at his feet, or like Yao Ming, one of the tallest Chinese basketball superstars standing out among us of average height. People used to call it "廿四层楼"(nian-si-ceng-lou) intimately and respectfully in Shanghainese, meaning "twenty-four-storey building". We stopped at the window on the third floor. The street outside was covered with golden and clear sunshine. Looking through a vast French window, I saw a big throng of people streaming and a long fleet of various vehicles bustling on Nanjing Road. Sometimes, the tramway car No.1 with lattice door would come by leisurely with melodious tinkles, dragging a long "tail" behind. My parent bought Little Friends as a birthday gift for me. A peaky old man with long white beard was smiling on one of its pages. I understood. It was the painter Qi Baishi who had won International Peace Award. But I did not have any sense of his greatness until several years later when I laid my eyes on the shrimps painted by him. They looked so vividly as if really moving in the invisible water. This touching moment in my childhood affects me all my life. So, I had an impression of a building called "Park Hotel", which was ranked the highest in the entire Asia for nearly half a century.

Close to Park Hotel stands another yellowish building on Nanjing Road, spaced by Huanghe Road, which seems to whisper intimate words with Park Hotel. On this building, there are rows of neon lights arranged in the form of sail highlighted on continuous tiles on the wall, shining six Chinese characters " 大光明电影院 " (da-guang-ming-dian-ying-yuan) on the one side and two English words "GRAND THEATRE" on the other. In the deep night, these neon signs sparkle proudly in the sky.

In the 1980s and 1990s, I was deep in the film industry. In the Grand Theatre were seated by many celebrities in Chinese filmdom like Bai Yang, Zhang Ruifang, Sun Daolin and Qinyi, and also Xie Jin and Qin Yigong later, who set the course for Chinese film industry, the kingdom on the screen. Here in the Grand Theatre, film artists like Zhang Yimou, Chen Kaige and Wu Ziniu who rose to great fame years later, also gathered together, talking shop and witnessing the memorable flourishing golden years of film industry gone by.

Later on, I worked in Shanghai Writers' Association -- the paradise I dreamed of in my childhood. In the beginning, I was told that our office was once the courtyard of W.S. LIEU, the king of matchstick. Several days later, someone else corrected it. It was the residence of K.S. LIEU, the younger brother of W.S. LIEU. Inlaid on the wall there is a long color glass window with painted landscape. The mystically brilliant light gives a sense of eminence to the main staircase circling up from the first floor to the third floor. The railing of the staircase has three letters "KSL" engraved on it, which is the abbreviation of K.S. LIEU to identify the owner of this residence. Seen from the top of the staircase, a beautiful curve spirals upward like a shell, with a floor on its foot paved with black and white tiles in a clear form as simple as a math equation. I was working in Room 204 in the southwest section of this villa, together with Ye Zhongyu, my tutor and Ye Xin, the poet, all my seniors, and my peers Zhao Changtian, Zong Fuxian and Ye Xin. On the south façade, there is a cast iron gate sandwiched by two ash tree wardrobes with a grand mirror and European patterns. Open the gate and you will see a garden in the yard evergreen with great vigor, oblivious of whatever happens in the world. The marble statue of Psyche (preserved by Mr. Li, the gardener serving Ba Jin, the renowned writer in Cultural Revolution) gracefully stretches her half-naked body, gazing far into the sky, with several babies playing around her, high above a pool of clear on the base. Sometimes, I would stand on the balcony alone or with Changtian where I could almost touch the Greek Ionic columns that support the whole villa. The ivy vines and tree leaves forced their way on the brown wall, stretching over toward me.. The east section of this villa is the editorial department of HARVEST, a literary bimonthly. Li Xiaolin, the daughter of Mr. Ba Jin, and her fellow editors were buried in the incoming manuscripts all the time. There is another balcony outside the office of the editorial department, right over the access to Shanghai Writers' Association for my colleagues. On the 40th Anniversary of Shanghai Writers' Association, in order to have a good show in the tele-film, Zhao Changtian, Lu Xing'er, Peng Ruigao, Zhu Lin and me sang at the top of our voices Young Friends, Let's Get Together, a famous Chinese pop song in the 1980s. Amusingly, over forty years old, we were all behaving like youngsters! Room 204 was the birthplace of all the stuffs of the establishment of Shanghai Writers' Association. No matter right or wrong, it is all gone together with thrilling and romantic stories of the owner of this villa. But now, only the villa stands here in the yard.

When I worked in Shanghai Municipal Bureau of Culture, I would visit a couple of old artists Mr. Sun Daolin and his wife Wang Wenjuan in any holiday. They lived on the fourth floor of a seven-storey high-rise at the crossing of Huaihai Road and Wukang Road. From the top, the building looks like a pressing iron. From the front, it looks like a great ship ready to set sail on the stormy sea. This building radiates an extraordinary sense of solemnity, grandeur and gracefulness to its surroundings. The terrace on the 7th floor offers an extensive view which seems to take in almost everything on earth. The entire building is like a symphony composed by Gustave Mahler, grand and orderly. The continuous gateway underneath the long arcade is like a resonant phrase coming out from the trumpet beat by beat. In the past, apart from Sun Daolin and Wang Wenjuan, many of Chinese popular movie stars like Zhao Dan, Wang Renmei, Shangguan Yunshu, Qin Yi and Zheng Junli enjoyed their "stay" in this building. With guidance of Sun Daolin and Wang Wenjuan, I was really touched by its charm and beauty. This building once had a powerful and resonant name: Normandie Apartments. Now, its name is too ordinary: Wukang Building. Furthermore, I wrote a commentary on Blue House. Coincidently, the prototype of Blue House is the "green house", or the residence of D.V. Woo, a tycoon of dyes. In the corridor of East China Hospital, I smelled the odor of solemn silence blended with soda water ….A wild geese disappears in the sky but with a resonant sound lingering in our mind. Here and there, the city echoes with the sound of a man. The sound is architecture, the music materialized on the ground.

I am sorry for my little knowledge. Not until many years later did I get to know that all these buildings are designed by the same master architect. He is L. E. Hudec, a renowned architect from Europe, whose nationality remains a mystery to us. We are not sure whether he is Hungarian or Slovakian. He has two homelands—Hungaria and Slovakia. It is such an architect who has left behind him nearly one hundred buildings in Shanghai, ranging from residences, cinemas, churches, hospitals to hotels, one third of which are listed as Shanghai Municipal Excellent Architectural Heritage. They are silent but brilliant, adding charm to the entire city. Up to now, no any other architect can be his match as far as his relationship with Shanghai is concerned. So, the architectural historians make a judgment that without L. E. Hudec, the history of Shanghai architecture will be totally different, or the history of modern Shanghai architecture has to be recompiled. This is a really rare case throughout the history of human's architecture.

In terms of the importance of an architect to a city, Hudec reminds me of Antonio Gaudi, a born genius of architecture for Barcelona in Spain, a city as great as Shanghai. Although the greatest modern artists in the 20th century such as Pablo Picasso, Joan Miro and Salvador Dali were somehow related to Barcelona, yet when I am walking on the streets of Barcelona, who touches me most is not anyone of them but the architect called "Gau-di". It seems that his incredible imagination full of beauty and elegance comes from the paradise high above and lands on the roofs, eaves, chimneys, walls, doors and windows on the streets of Barcelona. Casa Mila, Casa Batra, Parc Guell… like curling waves, surge towards me. And inside these buildings there are many seashell-like spirals and curves, oozing naturalistic mystery, fantasy and liveliness as well. Behold, the 170-meter high spire of Sagrada Familia stretches up, penetrating the blue sky the way the Tower of Babel was supposed to. Like a grand symphony with labyrinth structure, Sagrada Familia has been surrounded by scaffolds and cranes for more than a century. It is still "under construction" that seems never to be finished. Wandering in Barcelona, you are always in the arms of Gaudi. Similarly, in Shanghai, I, a boy from a poor family, was already embraced by Hudec in my innocent childhood.

What interests me most is that Gaudi and Hudec have different but equally marvelous lives. Hudec's legend is that he, a POW, escaped from prison, went across the Siberian plain covered with thick snow and finally settled down in Shanghai, a great city in the Far East. He, a homeless architect, left a lot of brilliant "houses" in this metropolis of Shanghai, and became the most famous architect in the history of Shanghai architecture. Then he must be very wealthy. But he was always an alien, awake or asleep. In the age of wars, he had no clear nationality and identity, and no homeland to settle down in. The very thought of this may have set his heart breaking. This pain deep in heart was really unspeakable. In 1918, he came to the Bund alone and then in 1947, he left Shanghai with all his family. He is like a shooting star in the night, giving out a magnificent glare at a sudden moment. He departed so determinedly that he designed no any other building except for the villa where he dwelled in! Finally, at the age of 65 he died of myocardial infarction in another foreign land.

Gaudi is eccentric and asocial. He was always dirty in the same ragged suit for three or even five years. He was never married. Finally, he was hit by the trolley car with colorful flags on its inauguration. The old ragged man was mistaken as a beggar. The moment he was taken to hospital, he had his final breath. Only one old lady identified him later. He died at the age of 74. All the citizens in Barcelona came out to attend his funeral.

Both Hudec and Gaudi contributed all their talents to a city. They, the great architects, deserve our respect and praise. We, the mediocre people are different in the extent of mediocrity, but, the fate of geniuses is always similar.

I am familiar with most of the painters invited to Love Home, Love Hudec. They live in this city, spending lots of time in their field trips on the streets and lanes, and work hard to paint the portrait of this city on the Xuan paper or canvas with their genuine innermost artistic language. This city has been their lover who dwells all the time in their heart. They are acutely sensitive to her gracious posture, her heartbeats, her expressions of being cold and warm, gloomy and shiny, and even her melancholy. All these are presented in their works as clear as the reality, and as remote and vague as the dream. They do love this city, love these houses, and love Hu - de - c, their designer as well.

As for me, finally I took off a day to visit Hudec's Residence sheltered in the lane of No. 129 Panyu Road. This Tudorbethan-country-style villa was built in 1930. In the background of a vast farmland in West Shanghai, the villa with half timbered frame and milky wall looked bright, pure and elegant, like a gracious lady standing in the breeze. It has suffered weathering for more than half a century.

Its doors and windows disengaged. Its walls cracked. Now, it looks dead, like an abandoned baby in the rain. This old villa, once so dignified, has decayed so miserably.

Yet, another alien lady from Hu'nan who had gone through much hardship in her business came across this villa and loved it at first sight. She would offer all that she has or even her life to the villa, like what an infatuated girl will do for her lover. She overcame many unthinkable difficulties in order to renovate this old villa, like rescuing a dying baby. She had worked with great efforts on every tiny corner of this villa for over 1000 days. Accompanied by her, I used to visit each room from the first floor to the third floor. None dwells there. In the great vacancy, the past days come back turbulently like silent tides in the moonlit. It seems that Hudec and his family are enjoying food around the table under the chandelier, their laughter echoing back to us. In the 21st century now, Hudec's Residence has regained its previous beauty and brilliance.

Night falls. Some light flickers in the dense wood. What a warm scene! The flying birds perch on the pointed ridge. They are the sculptures designed by Hudec, the owner of this villa. What is it really? Is it the soul of their owner that returns for a visit? Will he stay? Oh, the man who built houses for Shanghai.

张安朴
Zhang Anpu

1947年出生于上海。师从哈定、孟光。曾任解放日报美术编辑、摄影美术部主任、高级编辑。中国美术家协会会员，上海美术家协会理事。

40年创作了大批作品，多次获奖，并为中国人民邮政设计多套邮票。作品被中国美术馆、上海美术馆等机构收藏。

Zhang Anpu was born in Shanghai in 1947 and apprenticed to Ha Ding and Meng Guang. He was once the fine art editor, the director and a senior editor of Photography and Fine Art Department of Jiefang Daily. He is now a member of Shanghai Artists Association and a director of Shanghai Artists Association. He has created a great quantity of works for 40 years and won many prizes. He's designed many sets of post stamps for China Post. His works are collected in China Art Museum, Shanghai Art Museum and other institutes as well.

我是地道的"老上海",在这块土地上出生、读书、工作,对上海的老建筑有着特殊的情感。小时候与少年画友们去人民公园写生,不知不觉画了一张绿丛中的"沐尔堂"(是当时的称谓,现名为"沐恩堂"),结果被老师批评:那是"帝国主义分子"造的教堂,画它干什么!少年的我当时就有一个悬念,那个"帝国主义分子"是谁?数十年后,才知道这位建筑师叫邬达克,确是一位"帝国主义分子",因为他是匈牙利人,是"奥匈帝国"的一分子,一次世界大战后逃至东方上海的战俘……然而,正是这位"帝国主义分子"成为了上海城市建设的大功臣!出身于匈牙利建筑世家的邬达克把自己90%以上的作品留给了上海。他设计建造的国际饭店、大光明电影院、交通大学工程院、华东医院、沐恩堂、爱神别墅……都是上海的经典建筑,几乎全部列为"上海市优秀历史保护建筑"。尤其是国际饭店,其高度83米,是称雄半个世纪的"上海之巅"。

今天我们用画笔描绘邬达克建筑,是一次与邬达克先生的历史对话。因为邬达克先生不仅是一位杰出的建筑师,也是一位优秀的画家,我们彼此之间的艺术语言是相通的。当我们抚摸着邬达克故居的扶梯,仿佛能感觉到邬达克先生的温度,感受到他对建筑的理解,感悟到他对美的热爱,所以我们能产生"爱屋·及邬"的灵感,顿悟其中的内含,并深深地向邬达克先生致敬!

I am a "native Shanghainese in and out". I was born and educated in Shanghai, and I also work here. So, I have special feelings for its old buildings. When I was young, I went to People's Park to paint pictures with my peers. Unconsciously, I painted "Moore Memorial Church" (now named as Mu'en Church) in the green wood. Consequently, I was criticized by my teacher: That church was built by the "imperialist". Why painted it! But my curiosity was aroused. Who is that imperialist? Not until several decades later, did I get to know that its architect is L. E. Hudec, a real imperialist as he was Hungarian, a subject of Austro-Hungarian Empire, who fled to Shanghai from Concentration Camp in Russia after WWI. However, it is just this "imperialist" who has made great contributions to the construction of Shanghai! L. E. Hudec from a family of architects contributed 90% of his works to Shanghai. He designed Park Hotel, Grand Theatre, Engineering and Laboratory Building of Chiao Tung University, Huadong Hospital, Moore Memorial Church, Garden of Love, to mention just a few. All are classic buildings, and the overwhelming majority has been listed among "Shanghai Municipal Historic Heritage Architecture". Park Hotel, in particular, had been "the Peak of Shanghai" with its 83-meter-height for a half century.

Today, when we are painting his architecture, we are actually dialoguing with him for he is both an excellent architect and an outstanding artist, with whom we have the same language of art. When we touch the staircase in his residence, it seems that we can still feel his temperature, his passion for beauty and get to know his understanding of architecture. That gives us the inspiration of "Love Home, Love Hudec". That brings us to epiphany of its profound meanings. Let us pay a deep tribute to L. E. Hudec!

《大光明电影院 Grand Theatre》 26×20cm 钢笔水彩 watercolor with pen 2016

《西班牙式公寓 Spanish Apartments》 26×20cm 钢笔水彩 watercolor with pen 2016

《邬达克的家 Hudec's Home》 26×19cm 钢笔水彩 watercolor with pen 2016

《沐恩堂 Moore Memorial Church》 26×19cm 钢笔水彩 watercolor with pen 2016

《联合大楼 Union Building》 26×19cm 钢笔水彩 watercolor with pen 2016

《武康大楼 Wukang Building》 26×19cm 钢笔水彩 watercolor with pen 2016

《哥伦比亚住宅圈的别墅 Villas of Columbia Community》 19×26cm 钢笔水彩 watercolor with pen 2016

《英国乡村别墅 Britist Countryside Villa》 19×26cm 钢笔水彩 watercolor with pen 2016

《巨鹿路花园住宅 A Villa with Garden at Julu Road》 19×26cm 钢笔水彩 watercolor with pen 2016

《美丰银行大楼 American-Oriental Banking Corporation》 19×26cm 钢笔水彩 watercolor with pen 2016

陈 键
Chen Jian

1957年出生于上海。现为中国美术家协会会员,国家二级美术师。

曾获"全国小幅油画展"优秀作品奖,"全国写生画展"佳作奖,"第五届全国水彩粉画展"优秀作品奖,作品被上海世博会中国馆收藏。

Chen Jian was born in Shanghai in 1957. He is a member of China Artists Association and a National Accredited Second-Class Artist.

Awards: Excellent Works in "National Small Size Oil Painting Exhibition", Excellent Works in "National Life Sketching Exhibition", and Excellent Works in "The Fifth National Watercolor Painting Exhibition". His works have been collected in China Pavilion of Shanghai World Expo Park.

"在闹市、在角落，你都可能与他相逢"——这就是矗立在上海记忆里的邬达克。行走在上海西区的弄堂里，武康路、复兴路、永嘉路、安亭路……我要用我的画笔去守望和记忆"邬达克"。

"You will meet L. E. Hudec wherever you go, maybe in the busy downtown, or in a tranquil corner."–That is L. E. Hudec engraved in the memory of the Shanghainese. When I am walking along the lanes in West Shanghai, like Wukang Rd, Fuxing Rd, Yongjia Rd and Anting Rd, I will behold "Hudec" and I am trying to keep him alive in my mind with my painting brush.

《花园住宅——永嘉路563号 A Villa with Garden – No. 563 Yongjia Road》 22.5×30.5 cm 水彩 watercolor 2016

《花园住宅 —— 武康路 129 号 A Villa with Garden – No. 129 Wukang Road》 22.5×30.5 cm 水彩 watercolor 2016

《花园住宅——安亭路41弄18号 A Villa with Garden – Lane 41, Anting Road, No.18》 22.5×30.5 cm 水彩 watercolor 2016

《上海西区 West Shanghai》 22.5×30.5 cm 水彩 watercolor 2016

《爱司公寓 Estrella Apartments》 22.5×30.5 cm 水彩 watercolor 2016

《武康大楼 Wukang Building》 22.5×30.5 cm 水彩 watercolor 2016

《邬达克旧居 Hudec's Residence》 15×13.5cm 水彩 watercolor 2016

《邬达克旧居 Hudec's Residence》 15×13.5cm 水彩 watercolor 2016

《邬达克旧居 Hudec's Residence》 15×13.5cm 水彩 watercolor 2016

《邬达克旧居 Hudec's Residence》 15×13.5cm 水彩 watercolor 2016

贺寿昌

He Shouchang

1950年出生于上海。1977年毕业于上海戏剧学院。上海美术家协会会员,上海市政协书画院特聘画师,国家一级舞美设计师。作品多次参加展览,并为海内外机构及私人收藏。

He Shouchang was born in Shanghai in 1950 and graduated from Shanghai Theatre Academy in 1977. He is a member of Shanghai Artists Association, a senior painter of Shanghai Municipal CPPCC Painting and Calligraphy School, and is granted National Class-A Stage Art Designer as well. His works have been presented in many exhibitions and collected by many institutes and individuals from China and other countries.

把门打开
让那只鸽子飞进来!
他从西伯利亚的囚笼里飞出
他带着对生命的渴望在徘徊
这片土地接纳了他的灵魂和梦想
他把一个个百年时尚经典回报给上海

把门打开
让那只鸽子飞进来!
他从欧洲艺术殿堂里飞出
他衔着天边最美的那朵云彩
深夜的灯光映照出沐恩堂的圣火
他用宫殿用摩天大厦书写着心潮澎湃

把门打开
让那只鸽子飞进来!
他在蓝蓝的天蓝蓝的海上唱歌
他在鳞次栉比的楼群众盘桓
吸吮着东方女神的乳汁
他的心里溢满了对这座城市无限的爱

Open the door,
Let the dove glide in!
From the Siberian cage, he's just fled,
Circling around, cooing for life.
His soul, his dream, is accepted by this land.
In return of favor, he gives a time-tested fashion made by his hand.

Open the door,
Let the dove glide in!
From the palace of European art, he's flown away,
Carrying in his beak a cloud rosy-hued.
Lights late at night glamorize the holy Moore Memorial Hall.
His legend is written in palaces and high-rises, the best form of all.

Open the door,
Let the dove glide in!
He's singing above the blue sea and sky.
He's hovering among the clusters, low or high.
The Oriental Goddess breast-feeds him,
Whose heart wells up with endless adoration for Her.

回家 —— 这里有亲人和居所,这里是港湾和驿站

Back home — there are a family and a residence here, Or rather, a haven or a posthouse.

《回家 Back home》 69.5×49cm 纸上钢笔 pen-drawing on paper 2016

夜深 —— 柔情蜜意中，孕育着一件又一件新作

Deep in the night — new works are inspired one by one by affection and love.

《夜深 Deep in the Night》 69.5×49cm 纸上钢笔 pen-drawing on paper 2016

喂鸽 —— 远离了战争、远离了杀戮，他在这里播种生活

Feeding the doves — He was living peacefully here, far away from war and slaughter.

《喂鸽 Feeding the Doves》 69.5×49cm 纸上钢笔 pen-drawing on paper 2016

《春 Spring》 25×35cm 纸上钢笔 pen-drawing on paper 2013

《冬 Winter》 25×35cm 纸上钢笔 pen-drawing on paper 2013

《秋 Autumn》 35×25cm 纸上钢笔 pen-drawing on paper 2013

《夏 Summer》 35×25cm 纸上钢笔 pen-drawing on paper 2013

后记

与邬达克一样我也是在上海的异乡人，非建筑学专业，偶然的机缘来到上海，邂逅邬达克旧居后，才知道邬达克其人，了解并认识了邬达克留给上海的众多建筑。时光荏苒，邬达克成为我生命重心的日子已有六年，我不知道接下来与邬达克交集的生命旅程还有多长，这一切似乎是一场冥冥注定的缘分……

作为企业家，结缘邬达克是因为与长宁的"邬达克主题文化园区"设想。投资修缮邬达克旧居的过程，也是我对邬达克和邬达克留给上海的建筑遗产不断认识的过程，这份认识让我如履薄冰地修缮他在上海曾经最大的家——番禺路129号的邬达克旧居，唯恐辜负上海人民对邬达克的热爱。创建的邬达克纪念馆已累计接待中、外参观者达二十万人次。邬达克留给上海数量众多，种类分布广泛的建筑遗产，经岁月沉淀，许多已固化为经典的上海城市符号，成为上海人共同的城市记忆。行走在上海，无论你是否关注邬达克和他的建筑，你都可能与他的建筑不期而遇，或被他的建筑拥抱。

举办与邬达克和邬达克建筑相关的艺术创作展览，是我完成邬达克旧居修缮后的心愿之一。感谢文汇报高级编辑、市十一届政协常委、教科文卫体委员会副主任徐海清老师引荐，上海设计之都促进中心贺寿昌主任帮助，聘请到沪上知名艺术家李向阳老师担任策展人，"爱屋·及邬"来自李向阳老师的灵感。感谢中国文艺评论家协会副主席、上海美协理论委员会毛时安主任支持，出任"爱屋·及邬"纪念邬达克绘画雕塑邀请展及本书的艺术顾问。感谢毛冬华、杜海军、李乾煜、应海海、张安朴、陈键、洪健、贺寿昌（按姓名笔划）八位艺术家受邀参展。

还要感谢上海市文物局、上海市规划和国土资源管理局、上海市住房和城乡建设管理委员会、上海市长宁区人民政府的支持。同时，也要感谢上海市文物保护研究中心的鼎力合作，以及毛时安老师、徐海清老师、朱光老师的倾情撰文，还有邬达克文化发展中心的同仁们。是所有人的努力和付出，成就了"爱屋·及邬"艺术展和本书的出版。

邬达克也一定与上海有缘，不可思议的命运让这位才华横溢的战俘青年，在1918年来到他人生的福地——上海，他幸运的赶上了上海近代建筑繁荣崛起的时代，他的作品风格历经新古典主义、表现主义、装饰艺术派和现代建筑风格，仿佛上海建筑风格大全。邬达克在上海创造的传奇，折射出上海这个城市的包容与创新融合。今天我们以"爱屋·及邬"这场跨世纪、跨门类的艺术融合，希望唤起更多上海人对上海城市历史与未来的关注与思考。

刘素华
2016年10月18日

AFTERWORD

Just like L. E. Hudec I am not a native Shanghainese. And I was not a major of architecture. All by chance, I came to Shanghai and encountered Hudec's Residence, and then I began to know him and the buildings he's left here in Shanghai. Time flies. I have focused all my life on Hudec for six years. I am unsure how long I will be together with Hudec's heritage. It all seems that our relationship is predestined by the Almighty....

As an entrepreneur, I met Hudec by chance when Shanghai Changning District Government initiated the idea of "Hudec Cultural Theme Park" to attract investors to renovate Hudec's Residence. The more I learned about Hudec and his great contributions to Shanghai architectural heritage, the more cautious I turned to be in renovating his biggest house in Shanghai — Hudec's Residence at No. 129 Panyu Road for fear that we might fail the people of Shanghai and couldn't live up to their high expectations. After Hudec Memorial Hall was established and opened to the public, it has received 200 thousand presences of Chinese and foreigners as well. Hudec has left a great number of buildings of various styles, scattered in Shanghai, some of which have stood the test of time, and become classic symbols of Shanghai, engraved upon the memory of all its residents. Whether you care about Hudec and his architecture or not, you will definitely encounter his buildings or be surrounded by his buildings when you walk in the streets of Shanghai.

After I had renovated Hudec's Residence, a strong wish began to bud in my heart: to hold an art exhibition on the theme of Hudec and his architecture. Thanks to the introduction given by Mr. Xu Haiqing, Senior Editor of Wenhui Daily, a member of Shanghai 11th CPPCC, Deputy Director of Education, Sciences, Culture, Health Care and Sports Committee, and thanks to the help offered by Mr. He Shouchang, the director of Shanghai Capital of Design Promotion Center, we've successfully invited Mr. Li Xiangyang, an eminent artist in Shanghai to be the curator of this art exhibition. "Love Home, Love Hudec", the theme of this art exhibition originates from Li Xiangyang's artistic inspiration. My sincere gratitude also goes to Mr. Mao Shi'an, Vice-chairman of China Literary and Art Critics Association and Director of Theory Committee, Shanghai Artists Association. Mr. Mao Shi'an takes the role of art advisor for both "Love Home, Love Hudec"—a Painting and Sculpture Invitational Exhibition in Memory of L. E. Hudec and this album. And I am also indebted to the eight distinguished artists—Mao Donghua, Du Haijun, Li Qianyu, Ying Haihai, Zhang Anpu, Chen

Jian, Hong Jian and He Shouchang for their active participation in this exhibition.

I'd like to extend my heartfelt gratitude to Shanghai Municipal Administration for Protection of Cultural Relics, Shanghai Municipal Administration of Urban Planning and State Land Resource, Shanghai Municipal Housing and Urban and Rural Construction Committee and Shanghai Changning District People's Government for their support. Meanwhile, I greatly appreciate the cooperation with Shanghai Municipal Center for the Protection and Study of Cultural Relics, the articles written by Mr. Mao Shi'an, Mr. Xu Haiqing and Mr. Zhu Guang and the support offered by my colleagues in Hudec Cultural Development Center. It is due to all your great efforts that we are able to hold this exhibition and publish this album.

The relationship between Hudec and Shanghai must be predestined. It is the inconceivable Goddess of Fate who guided him, a prisoner of war full of talent to Shanghai, the land of fortune in 1918, where he caught up with the flourishing period of architecture in modern Shanghai. His works vary from Neo-classicism, Expressionism, Art Deco all the way to Modern Architecture, making an encyclopedia of architecture in Shanghai. The legend of L. E. Hudec created in Shanghai mirrors an integration of two characteristics of Shanghai—openness and innovation. Today, we hope to inspire more people to rethink and care about the past and the future of Shanghai by holding "Love Home , Love Hudec", a cross-century exhibition to converge three forms of art—architecture, painting and sculpture.

By Liu Suhua
On Oct. 18th, 2016

图书在版编目(CIP)数据

爱屋·及邬 纪念邬达克绘画雕塑邀请展:Love Home Love Hudec Painting and Sculpture Invitational Exhibition in Memory of L. E. Hudec/李向阳主编.—上海:上海远东出版社,2016
ISBN 978-7-5476-1212-5

Ⅰ.①爱… Ⅱ.①李… Ⅲ.①建筑设计—作品集—匈牙利—现代
Ⅳ.①TU206

中国版本图书馆CIP数据核字(2016)第271261号

爱屋·及邬 纪念邬达克绘画雕塑邀请展
Love Home Love Hudec
Painting and Sculpture Invitational Exhibition in Memory of L. E. Hudec
李向阳　主编
责任编辑/贺　寅　装帧设计/曹景宇

出版:上海世纪出版股份有限公司远东出版社
地址:中国上海市钦州南路81号
邮编:200235
网址:www.ydbook.com
发行:新华书店　上海远东出版社
　　　上海世纪出版股份有限公司发行中心
印刷:上海昌鑫龙印务有限公司
装订:上海昌鑫龙印务有限公司

开本:889×1194　1/16　印张:10.25　插页:2　字数:150千字
2016年12月第1版　2016年12月第1次印刷

ISBN 978-7-5476-1212-5/TU·106
定价:128.00元

版权所有　盗版必究(举报电话:62347733)
如发生质量问题,读者可向工厂调换。
零售、邮购电话:021-62347733-8538